SpringerBriefs in Applied Sciences and Technology

Computational Mechanics

T0238965

For further volumes:
http://www.springer.com/series/8886

Sergey Alexandrov

Upper Bound Limit Load Solutions for Welded Joints with Cracks

 Springer

Sergey Alexandrov
A.Yu. Ishlinskii Institute for Problems in Mechanics
Russian Academy of Sciences
Prospect Vernadskogo 101-1
Moscow
Russia 119526

ISSN 2191-530X   ISSN 2191-5318   (electronic)
ISBN 978-3-642-29233-0   ISBN 978-3-642-29234-7   (eBook)
DOI 10.1007/978-3-642-29234-7
Springer Heidelberg New York Dordrecht London

Library of Congress Control Number: 2012935677

Printed on acid-free paper

Springer is part of Springer Science+Business Media (www.springer.com)

Preface

This monograph concerns with application of the upper bound theorem to finding the limit load for welded structures including structures with cracks. The presentation of the introductory material and the theoretical developments appear in a text of six chapters. The topics chosen are primarily of interest to engineers as postgraduates and practitioners but they should also serve to capture a readership from among applied mathematicians. The monograph provides both a collection of limit load solutions for welded structures and a description of general approaches to finding the limit load for a class of structures. Many solutions are represented by formulae. Such solutions are immediately ready for practical use. Other solutions are illustrated by diagrams. These diagrams demonstrate most important tendencies in solutions behavior. It is however evident that they cannot be used for practical calculation of the limit load. Therefore, most of such solutions are described in great detail, including possible difficulties with application of numerical methods, and quantitative results can be easily reproduced. In most cases, numerical techniques are only necessary to evaluate integrals and minimize functions of one variable. As a rule, approximations of solutions by elementary functions are not given in the monograph. Although such approximations are widely used in the literature, it is believed that they are not efficient when there are several essential input parameters. For reasons of space, the main focus is on various highly undermatched tensile specimens, though undermatched and overmatched cases are briefly discussed as well.

Among the topics that are either new or presented in greater detail than would be found in similar texts are the following:

1. An approach to modifying upper bound solutions for a class of structures with no crack to account for the presence of a crack.
2. An approach to using singular velocity fields for constructing accurate upper bound solutions for highly undermatched joints.
3. The effect of the thickness of specimens on the limit load.
4. The effect of plastic anisotropy on the limit load.

5. A discussion of difficulties with application of numerical techniques in conjunction with simple kinematically admissible velocity fields.

Chapter 1 concerns with the upper bound theorem for rigid perfectly plastic materials. A formal proof is not given because it can be found in any text on plasticity theory. Instead, original and efficient approaches to finding upper bound limit load solutions for welded joints with and with no cracks are introduced and explained. These approaches are used in subsequent chapters.

Chapters 2–4 deal with highly undermatched specimens subject to tension. Firstly, in Chap. 2, two solutions for the center cracked specimen under plane strain conditions are presented. Each of these solutions illustrates one of the general approaches introduced in Chap. 1. The solutions found are generalized to scarf joint specimens in Chap. 3, also under plane strain conditions. Axisymmetric solutions are given in Chap. 4.

In Chap. 5 two solutions for pure bending of highly undermatched panels under plane strain conditions are discussed. One of these solutions is based on an exact solution of plasticity theory. The other solution is obtained with the use of one of the universal methods proposed in Chap. 1. Comparison of the solutions determines the ranges in parametric space where each of them should be adopted.

Chapter 6 includes a brief discussion of several topics. Firstly, the effect of the thickness of panels on the limit load is illustrated. To this end, the solution for the center cracked specimen presented in Chap. 2 is compared to a new three-dimensional solution. The effect of the mis-match ratio is discussed next. Solutions for the undermatched and overmatched center cracked specimen are given and the definition for the highly undermatched case is clarified. Finally, it is shown that the effect of plastic anisotropy on the limit load is very significant and this material property should not be ignored in the development of flaw assessment procedures.

Contents

Symbols

The intention within the various theoretical developments given in this monograph has been to define each new symbol where it first appears in the text. A number of symbols are introduced in the abstract to individual chapters. These symbols re-appear consistently throughout the chapter. In this regards each chapter should be treated as self-contained in its symbol content. There are, however, certain symbols that re-appear consistently throughout the text. These symbols are given in the following list.

F	Tensile force
F_u	Upper bound on F
f_u	Dimensionless representation of F_u
G	Bending moment
G_u	Upper bound on G
g_u	Dimensionless representation of G_u
u_x, u_y, u_z	Components of kinematically admissible velocity field in Cartesian coordinates
$|[u_\tau]|$	Amount of velocity jump
x, y, z	Cartesian coordinates
ζ_{eq}	Equivalent strain rate found using kinematically admissible velocity field
$\zeta_{xx}, \zeta_{yy}, \zeta_{zz},$ $\zeta_{xy}, \zeta_{xz}, \zeta_{yz}$	Components of kinematically admissible strain rate field in Cartesian coordinates
ξ_{eq}	Equivalent strain rate
σ_0	Yield stress in tension
$\mathbf{n}$	Unit normal vector
$\mathbf{U}$	Velocity vector in rigid zone
$\mathbf{u}$	Velocity vector in plastic zone

Chapter 1
Upper Bound Theorem

Plastic limit analysis is a convenient tool to find approximate solutions of boundary value problems. In general, this analysis is based on two principles associated with the lower bound and upper bound theorems. The latter is used in the present monograph to estimate the limit load for welded structures with and with no crack. A proof of the upper bound theorem for a wide class of material models has been given by Hill (1956). The only reliable output of upper bound solutions is the load required to initiate the process of plastic deformation. Any upper bound limit load is higher than or equal to the actual load. This statement becomes more complicated in the case of multiple load parameters. Upper bound solutions are not unique and their accuracy significantly depends on the kinematically admissible velocity field chosen. Therefore, the development of methods for constructing kinematically admissible velocity fields accounting for some mathematical features of real velocity fields is of great importance for successful applications of the method. In addition to the methods described in the present monograph, original approaches have been proposed in Sawczuk and Hodge (1968), Wilson (1977), and Yeh and Yang (1996) among others.

Upper bound solutions overestimate the load carrying capacity of structures. Therefore, such solutions may be associated with one possible failure mechanism of structures. However, a more important application of limit load solutions to structures with cracks is that the limit load is an essential input parameter of flaw assessment procedures. Therefore, the accuracy of the limit load found has a great effect of the accuracy of predictions made with the use of these procedures. A review of flaw assessment procedures can be found in Zerbst et al. (2000). Reviews of limit load solutions for structures with cracks are available in Miller (1988) and Alexandrov (2011).

S. Alexandrov, *Upper Bound Limit Load Solutions for Welded Joints with Cracks*, SpringerBriefs in Computational Mechanics, DOI: 10.1007/978-3-642-29234-7_1, © The Author(s) 2012

1.1 Basic Assumptions and Equations

The present monograph concerns with rigid perfectly plastic material. This means that the elastic portion of the strain rate tensor is neglected and all yield stresses are material constants. It is worthy of note here that the former assumption is not essential since the limit load is independent of elastic properties (Drucker et al. 1952). Strain hardening has no effect on the limit load as well, as long as the initial configuration is of concern. The constitutive equations of the model chosen consist of the yield criterion and its associated flow rule. A great account on this model is given in Hill (1950). Extension of the theory to piece-wise homogeneous materials, as it is required for welded structures, is in general straightforward. It is presented in Rychlewski (1966) under plane strain conditions. The Mises yield criterion is adopted throughout this monograph (except anisotropic solutions given in Chap. 6). This criterion can be written in the form

$$\sqrt{\frac{3}{2}\tau_{ij}\tau_{ij}} = \sigma_0. \tag{1.1}$$

Here and in what follows the summation convention, according to which a recurring letter suffix indicates that the sum must be formed of all terms obtainable by assigning to the suffix the values 1, 2, and 3, is adopted. Similarly, in a quantity containing two repeated suffixes, say i and j, the summation must be carried out for all values 1, 2, 3 of both i and j. Also, $\tau_{ii} = \sigma_{ij} - \sigma\delta_{ij}$, σ_{ij} are the components of the stress tensor, $\sigma = \sigma_{ij}\delta_{ij}/3$, δ_{ij} is the Kroneker symbol, and σ_0 is the yield stress in tension.

A proof of the upper bound theorem can be found, for example, in Hill (1950). When the yield criterion (1.1) is adopted, the theorem reads

$$\iint_{S_v} (t_i v_i)\,dS \leq \sigma_0 \iiint_V \zeta_{eq}\,dV - \iint_{S_f} (t_i u_i)\,dS + \frac{\sigma_0}{\sqrt{3}}\iint_{S_d} \|[u_\tau]\|\,dS \tag{1.2}$$

where V is the volume of material loaded by prescribed external stresses t_i over a part S_f of its surface, and by prescribed velocities over the remainder S_v. Also, v_i are the components of the real velocity vector, u_i are the components of a kinematically admissible velocity vector, ζ_{eq} is the equivalent strain rate, and $\|[u_\tau]\|$ is the amount of velocity jump across the velocity discontinuity surface S_d. The velocity component u_τ should be found using the kinematically admissible velocity field. The normal velocity must be continuous across the velocity discontinuity surface. The equivalent strain rate involved in Eq. (1.2) is defined by

$$\zeta_{eq} = \sqrt{\frac{2}{3}\zeta_{ij}\zeta_{ij}} \tag{1.3}$$

where ζ_{ij} are the strain rate components. These components should be found using the kinematically admissible velocity field u_i. The left hand side of the inequality

(1.2) can be evaluated using any kinematically admissible velocity field. The boundary value problems considered in the present monograph contain just one unknown load. Therefore, this load can be evaluated using Eq. (1.2). Useful results for boundary value problems containing several independent load parameters can be found in Hodge and Sun (1968). The second integral on the right hand side of Eq. (1.2) usually include traction free and frictional surfaces. There are no frictional surfaces in the boundary value problems considered in the present monograph. Moreover, $t_i = 0$ over any traction free surface. Therefore, the second integral on the right hand side of Eq. (1.2) vanishes and the inequality simplifies to

$$
\iint\limits_{S_v} (t_i v_i)\, dS \leq \sigma_0 \iiint\limits_{V} \zeta_{eq}\, dV + \frac{\sigma_0}{\sqrt{3}} \iint\limits_{S_d} |[u_\tau]|\, dS. \tag{1.4}
$$

Many simple upper bound solutions are obtained by assuming that the kinematically admissible velocity field consists of rigid blocks. In this case $\zeta_{eq} = 0$ in V and the inequality (1.4) further simplifies to

$$
\iint\limits_{S_v} (t_i v_i)\, dS \leq \frac{\sigma_0}{\sqrt{3}} \iint\limits_{S_d} |[u_\tau]|\, dS. \tag{1.5}
$$

The physical meaning of the left hand side of Eq. (1.4) is the rate at which external forces do work. The physical meaning of the first term on its right hand side is the rate of work dissipation in plastic zones and the physical meaning of the second term is the rate of work dissipation at velocity discontinuity surfaces.

By definition, any velocity field satisfying the velocity boundary conditions and the equation of incompressibility is a kinematically admissible velocity field. The equation of incompressibity can be written as

$$
\zeta_{ii} = 0. \tag{1.6}
$$

The condition that the normal velocity must be continuous across velocity discontinuity surfaces can be represented as the following scalar product of two vectors.

$$
(\mathbf{u_1} - \mathbf{u_2}) \cdot \mathbf{n} = 0 \tag{1.7}
$$

where $\mathbf{u_1}$ and $\mathbf{u_2}$ are the velocity vectors on sides 1 and 2 of the velocity discontinuity surface, respectively, and $\mathbf{n}$ is the unit normal vector to this surface. Then, the amount of velocity jump can be found as

$$
|[u_\tau]| = \sqrt{(\mathbf{u_1} - \mathbf{u_2}) \cdot (\mathbf{u_1} - \mathbf{u_2})}. \tag{1.8}
$$

It is obvious that it is possible to propose any number of kinematically admissible velocity fields for any problem. If a kinematically admissible velocity field coincides with the real velocity field then Eq. (1.2) gives the exact value of the limit load. In most cases however, kinematically admissible velocity fields result in upper bounds on the limit load. If a kinematically admissible velocity field chosen contains no free parameters, its substitution into Eq. (1.2) immediately gives an upper bound on the limit load. A better prediction can be indeed achieved when a kinematically admissible velocity field contains free parameters. In such cases, substituting this velocity field into Eq. (1.2) transforms the functional on its right hand side into a function of one or several variables. It is obvious from the structure of the inequality (1.2) that its right hand side should be minimized with respect to these variables to find the best upper bound limit load based on the kinematically admissible velocity field chosen. In general, there are two main approaches to handle this problem: (1) numerical methods based on finite element approximation, and (2) analytical and semi-analytical methods. The former have been developed, for example, in Chang and Bramley (2000) and Bramley (2001). The original approaches have been developed for metal forming simulation. However, they are also applicable for structural analysis. The present monograph is devoted to analytical and semi-analytical methods of plastic limit analysis. Such methods are very useful for engineering applications (Schwalbe 2010).

There is the companion theorem to the upper bound theorem to find the lower bound limit load (see, for example, Hill 1950). This theorem is not considered in the present monograph and all limit loads in subsequent chapters should be understood as the upper bound limit loads.

1.2 Vicinity of the Bi-Material Interface

Even though any kinematically admissible velocity field results in an upper bound on the limit load, it is advantageous to choose a kinematically admissible velocity field which takes into account behavior of the real velocity field that must exist near some surfaces. For welded joints, the bi-material interface is such a surface when it is also a velocity discontinuity surface. The latter is a typical situation for highly undermatched joints. In such joints, the weld is much softer than the base material and plastic deformation is localized within the weld whereas the base material is rigid. Such compositions of materials are of practical interest (Hao et al. 1997). According to the general theory, the shear stress at any velocity discontinuity surface coinciding with a bi-material surface must be equal to the shear yield stress of the softer material. In particular, Rychlewski (1966) has considered such a distribution of stresses in the case of plane strain deformation. Alexandrov and Richmond (2001) have shown that the real velocity field must be singular near surfaces on which the shear stress is equal to the shear yield stress (there are exceptions to this rule but those are not significant in most cases of practical interest). They have also proposed a conceptual approach to use this property of the real velocity field in plastic limit analysis of structures (Alexandrov and Richmond 2000). The main result obtained in Alexandrov and Richmond (2001) can be represented by the following equation

$$\xi_{eq} = D/\sqrt{s} + o(1/\sqrt{s}), \quad s \to 0 \tag{1.9}$$

where ξ_{eq} is the equivalent strain rate found using the real velocity field, D is the strain rate intensity factor, and s is the normal distance to the velocity discontinuity surface. The strain rate intensity factor is independent of s. It has been proposed in Alexandrov and Richmond (2001) to choose kinematically admissible velocity fields such that the distribution of ζ_{eq} satisfies the inverse square root rule shown in Eq. (1.9). When such a kinematically admissible velocity field is chosen, the stress boundary condition at the corresponding velocity discontinuity surface is automatically satisfied. A comprehensive comparative study of the effect of various kinametically admissible fields on the limit load for axisymmetric extrusion of rods has been completed in Alexandrov et al. (2001) and it has been shown that the one satisfying Eq. (1.9) gives the best upper bound as compared to other kinematically admissible velocity fields of the same level of complexity. Substituting Eq. (1.9) into (1.2) where ζ_{eq} should be replaced with ξ_{eq} shows that the volume integral is improper. Even though it is easy to demonstrate its convergence, such singular behavior can cause difficulties with numerical integration. In the case of methods used in the present monograph, these difficulties can be overcome by changing the variable of integration or/and expanding functions in a series in the vicinity of velocity discontinuity surfaces. These techniques will be demonstrated in subsequent chapters.

1.3 Effect of Cracks

There is a class of structures for which the effect of a crack on the limit load can be accounted for by a simple formula if the limit load for the structure with no crack is available (Alexandrov and Kocak 2008). The formula is exact (i.e. this formula transforms the exact limit load for a structure with no crack into the exact limit load for the structure with a crack).

A typical welded structure of the class under consideration and the directions of the axes of Cartesian coordinates (x, y, z) are illustrated in Fig. 1.1. It is obvious that the corresponding structure made of homogeneous material is obtained as a particular case. The structure must have a plane of symmetry orthogonal to the line of action of tensile forces F. With no loss of generality, it is possible to assume that the equation for this plane is $z = 0$. The structure contains a through crack of length $2a$ in the plane of symmetry. In order to precisely define the class of structures under consideration, it is necessary to consider a particular case of the structure containing no crack ($a = 0$). It is required that all cross-sections $y = $ constant are identical. This means in particular that the structure has a plane of symmetry orthogonal to the y-axis. It is convenient to assume, with no loss of generality, that the equation for this plane is $y = 0$. Then, two boundaries of the structure are determined by the equations $y = \pm 2W$ where $2W$ is the width of the structure.

Fig. 1.1 Geometry of the
structure under
consideration—notation

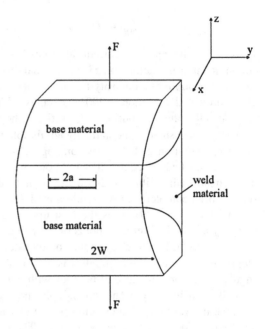

Mechanical properties may vary arbitrarily in the x-direction, must be symmetric relative to the plane $z = 0$ and must be identical for all cross-sections $y = $ constant. It is assumed that constraints imposed on the structure force its identical rigid parts located above and below the plane of symmetry $z = 0$ move along the z-axis in the opposite directions with the same velocity. In addition, for the general case of the structure with a crack, its tips are determined by the equations $y = \pm a$.

Using dimensional analysis it is possible to write

$$\frac{F_u^{(0)}}{4BW\sigma_0} = \Omega(w) \tag{1.10}$$

where $F_u^{(0)}$ is the limit load for the structure with no crack, $2B$ is the thickness of the structure at $z = 0$, $w = W/L$, L is a reference length, and $\Omega(w)$ is a function of w found for the structure with no crack by means of the upper bound theorem. In addition to w, this function may depend on other non-dimensional arguments. These other arguments, however, do not contain W. In the simplest case $\Omega(w)$ is independent of any argument. The argument w is explicitly shown in the notation for $\Omega(w)$ for further convenience.

A typical shape of the plastic and rigid zones in the domain $y \geq 0$ and $z \geq 0$ (the remainder of the cross-section is not considered because of symmetry) is illustrated in Fig. 1.2. The rigid zone moves with velocity U along the z-axis. The magnitude of this velocity is of course immaterial. The main assumption concerning the shape of the plastic zone is that it is entirely located on the right to plane Σ. This plane is orthogonal to the y-axis and contains the crack tip. For the

Fig. 1.2 Typical location of
velocity discontinuity surface
and plastic zone

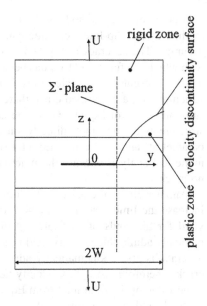

structure with no crack the plane Σ coincides with the plane $y = 0$. Let E_0 be the
rate of work dissipation in the plastic zone, including possible velocity disconti-
nuities, for the specimen with no crack. Then, since there are four identical plastic
zones and two identical tensile forces act, it follows from the principle of virtual
work rate that $4E_0 = 2UF_u^{(0)}$. Using Eq. (1.10) gives

$$E_0 = 2U\sigma_0 BW\Omega(w). \tag{1.11}$$

Taking into account geometrical and physical properties of the class of structures under consideration it is possible to conclude that the only difference in the
rate of work dissipation in the plastic zone shown in Fig. 1.2 for different a-values
results from the size of this zone at $z = 0$. This size is equal to W for the structure
with no crack and $W - a$ for the structure with the crack. Therefore, using
Eq. (1.11) the rate of work dissipation in the plastic zone for the cracked structure
can be found as

$$E_c = 2U\sigma_0 B(W - a)\Omega(w_c) \tag{1.12}$$

where $w_c = (W - a)/L$. Using the principle of virtual work rate and Eq. (1.12)
the upper bound on the limit load for the cracked structure can be obtained in the
form

$$\frac{F_u^{(c)}}{4BW\sigma_0} = \left(1 - \frac{a}{W}\right)\Omega(w_c). \tag{1.13}$$

Thus, once the function $\Omega(w)$ has been found, the limit load can be immediately
determined by means of Eq. (1.13) with no difficulty. Note that there is no

restriction on the method used to find $\Omega(w)$. In particular, the finite element method or the slip-line technique can be adopted to determine $\Omega(w)$ with a high accuracy. The accuracy of Eq. (1.13) for the limit load is the same as the accuracy with which the function $\Omega(w)$ has been determined. Of course, an advantage of the method presented is revealed only if parametric analysis is of concern. For any structure with a prescribed crack there is no difference between finding $\Omega(w)$ and subsequent use of Eq. (1.13) on one hand and determining the limit load for the structure with the crack directly on the other hand. However, in the case of parametric analysis using Eq. (1.13) reduces the number of parameters in numerical analysis because the function $\Omega(w)$ can be found for the structure with no crack.

Another effect of cracks on the limit load is that the presence of a crack cannot increase the limit load as compared to the structure with no crack. This effect is valid for all kinds of structures. A formal proof is straightforward. Any kinematically admissible velocity field used to find the limit load for the structure with no crack is also a kinematically admissible velocity field for the structure with a crack. Therefore, using this velocity field for the cracked structure one can get the same value of the limit load from Eq. (1.2). Of course, in most cases the presence of cracks reduces the limit load.

References

S. Alexandrov, Plastic limit load solutions for highly undermatched welded joints, in *Welding: Processes, Quality, and Applications*, ed. by R.J. Klein (Nova Science Publisher, Hauppauge, 2011)

S. Alexandrov, M. Kocak, Effect of three-dimensional deformation on the limit load of highly weld strength undermatched specimens under tension. Proc. Inst. Mech. Eng. Part C: J. Mech. Eng. Sci. **222**, 107–115 (2008)

S. Alexandrov, O. Richmond, On estimating the tensile strength of an adhesive plastic layer of arbitrary simply connected contour. Int. J. Solids Struct. **37**, 669–686 (2000)

S. Alexandrov, O. Richmond, Singular plastic flow fields near surfaces of maximum friction stress. Int. J. Non-Linear Mech. **36**, 1–11 (2001)

S. Alexandrov, G. Mishuris, W. Mishuris et al., On the dead zone formation and limit analysis in axially symmetric extrusion. Int. J. Mech. Sci. **43**, 367–379 (2001)

A.N. Bramley, UBET and TEUBA: fast methods for forging simulation and perform design. J. Mater. Process. Technol. **116**, 62–66 (2001)

C.C. Chang, A.N. Bramley, Forging perform design using a reverse simulation approach with the upper bound finite element procedure. Proc. Inst. Mech. Eng. Part C: J. Mech. Eng. Sci. **214**, 127–136 (2000)

D.C. Drucker, W. Prager, H.J. Greenberg, Extended limit design theorems for continuous media. Q. J. Appl. Mech. **9**, 381–389 (1952)

S. Hao, A. Cornec, K.-H. Schwalbe, Plastic stress-strain fields and limit loads of a plane strain cracked tensile panel with a mismatched welded joint. Int. J. Solids Struct. **34**, 297–326 (1997)

R. Hill, *The Mathematical Theory of Plasticity* (Clarendon Press, Oxford, 1950)

R. Hill, New horizons in the mechanics of solids. J. Mech. Phys. Solids **5**, 66–74 (1956)

P.G. Hodge Jr, C.-K. Sun, General properties of yield-point load surfaces. Trans. ASME. J. Appl. Mech. **35**, 107–110 (1968)

A.G. Miller, Review of limit loads of structures containing defects. Int. J. Press. Vessels Pip. **32**, 197–327 (1988)

J. Rychlewski, Plain plastic strain for jump of non-homogeneity. Int. J. Non-Linear Mech. **1**, 57–78 (1966)

A. Sawczuk, P.G. Hodge Jr, Limit analysis and yield-line theory. Trans. ASME J. Appl. Mech. **35**, 357–362 (1968)

K.-H. Schwalbe, On the beauty of analytical models for fatigue crack propagation and fracture—a personal historical review. J. ASTM Int. **7**(8), Paper ID 102713 (2010)

W.R.D. Wilson, A simple upper-bound method for axisymmetric metal forming problems. Int. J. Mech. Sci. **19**, 103–112 (1977)

W.-C. Yeh, Y.-S. Yang, A variational upper-bound method for plane strain problems. Trans. ASME J. Manuf. Sci. Eng. **118**, 301–309 (1996)

U. Zerbst, R.A. Ainsworth, K.-H. Schwalbe, Basic principles of analytic flaw assessment methods. Int. J. Press. Vessels Pip. **77**, 855–867 (2000)

Chapter 2
Plane Strain Solutions for Highly Undermatched Tensile Specimens

The specimens considered in this chapter are welded plates with the weld orientation orthogonal to the line of action of tensile forces applied. A crack is entirely located in the weld. Edge cracks are excluded from consideration. The width of the plate is denoted by $2W$, its thickness by $2B$, the thickness of the weld by $2H$, and the length of the crack by $2a$ (except the last solution of this chapter which deals with cracks of arbitrary shape in the plane of flow). Since plane strain solutions are of concern in the present chapter, integration in the thickness direction in volume and surfaces integrals involved in Eq. (1.4) is replaced with the multiplier $2B$ without any further explanation. For the same reason, the term "velocity discontinuity surface" is replaced with the term "velocity discontinuity curve (or line)". The latter refers to curves (lines) in the plane of flow. Base material is supposed to be rigid.

2.1 Center Cracked Specimen

The geometry of the specimen and the directions of the axes of Cartesian coordinates (x, y) are illustrated in Fig. 2.1. The specimen is loaded by two equal forces F whose magnitude at plastic collapse should be evaluated. A numerical slip-line solution for such specimens has been proposed in Hao et al. (1997). It is evident that Eq. (1.13) is valid for the specimen under consideration. It is possible to assume that $L \equiv H$. It is first necessary to determine the limit load for the specimen with no crack, $a = 0$. It is convenient to choose the origin of the Cartesian coordinate system at the intersection of the axes of symmetry of the specimen. Then, it is sufficient to find the solution in the domain $x \geq 0$ and $y \geq 0$. Let u_x be the velocity component in the x-direction and u_y in the y-direction. The blocks of rigid base material move with a velocity U along the y-axis in the opposite directions. The velocity boundary conditions are

S. Alexandrov, *Upper Bound Limit Load Solutions for Welded Joints with Cracks*, SpringerBriefs in Computational Mechanics, DOI: 10.1007/978-3-642-29234-7_2, © The Author(s) 2012

Fig. 2.1 Geometry of the specimen under consideration–notation

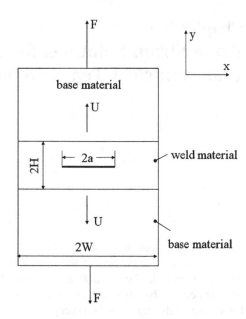

$$u_y = 0 \tag{2.1}$$

for $y = 0$,

$$u_x = 0 \tag{2.2}$$

for $x = 0$,

$$u_y = U \tag{2.3}$$

for $y = H$, and

$$u_x = 0 \tag{2.4}$$

for $y = H$.

The boundary conditions (2.1) to (2.4) show that the present problem is equivalent to the problem of plane strain compression of a plastic layer between two parallel, rigid plates if the maximum friction law is assumed at the surface of plates (the difference in the sense of u_y at $y = H$ is immaterial for pressure-independent materials). An approximate stress solution for the latter has been obtained by Prandtl (1923) and an approximate velocity solution by Hill (1950). Hill (1950) has also found an accurate slip-line solution. Using this solution the force can be approximated by

$$\frac{F_u^{(0)}}{4\sigma_0 BW} = \frac{1}{2\sqrt{3}}\left(3 + \frac{W}{H}\right). \tag{2.5}$$

Fig. 2.2 Configuration of
plastic and rigid zones from
the slip-line solution

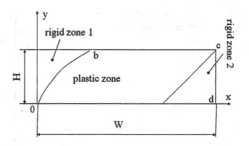

This approximation is valid for $W/H \geq 1$. A schematic diagram showing the configuration of plastic and rigid zones which follows from the slip-line solution is presented in Fig. 2.2 for one quarter of the weld. There are two velocity discontinuity curves, $0b$ and bc, and the solution satisfies Eq.(1.9) in the vicinity of line bc. It follows from Eqs. (1.10) and (2.5) that

$$\Omega(w) = \frac{(3+w)}{2\sqrt{3}}. \tag{2.6}$$

Therefore, substituting Eq. (2.6) into Eq. (1.13) gives the limit load for the cracked specimen in the form

$$f_u = \frac{F_u}{4\sigma_0 BW} = \frac{1}{2\sqrt{3}}\left(1 - \frac{a}{W}\right)\left(3 + \frac{W-a}{H}\right). \tag{2.7}$$

The restriction $W/H \geq 1$ transforms to $(W-a)/H \geq 1$. Thus the solution is not valid for large cracks. The solution for this special case is trivial and is available in the literature (see, for example, Kim and Schwalbe 2001a).

The derivation that has led to the solution (2.7) is an illustration of using Eq. (1.13) in conjunction with numerical solutions. The problem under consideration is also very suitable for illustrating the use of Eq. (1.9) in upper bound solutions. Therefore, even though it is not realistic to obtain a better result than that given by Eq. (2.7), such a solution is provided below.

A general approach to construct singular kinematically admissible velocity fields for plastic layers subject to tensile loading has been proposed in Alexandrov and Richmond (2000). The starting point of this approach is the representation of velocity components tangent to the bi-material interface in such a form that Eq. (1.9) is automatically satisfied. Then, the solution to the equation of incompressibility (1.6) gives the axial velocity in rather a complicated form. Moreover, the kinematically admissible velocity field proposed in Alexandrov and Richmond (2000) contains no rigid zone near the axis of symmetry (such as rigid zone 1 in Fig. 2.2). Therefore, a slightly different approach is developed here. It is convenient to introduce the following dimensionless quantities.

$$\frac{y}{H} = \eta, \quad \frac{x}{W} = \varsigma, \quad \frac{H}{W} = h. \tag{2.8}$$

Fig. 2.3 General structure of
the kinematically admissible
velocity field

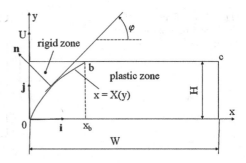

The starting point of the present approach is a linear through-thickness distribution of the axial velocity, which is quite reasonable because the thickness of the layer is small as compared to its width. Then, the boundary conditions (2.1) and (2.3) along with Eq. (2.8) require that

$$\frac{u_y}{U} = \eta. \tag{2.9}$$

The equation of incompressibility (1.6) in the case under consideration reduces to

$$h\frac{\partial u_x}{\partial \varsigma} + \frac{\partial u_y}{\partial \eta} = 0. \tag{2.10}$$

Substituting Eq. (2.9) into Eq. (2.10) and integrating give

$$\frac{u_x}{U} = -\frac{\varsigma}{h} + g(\eta) \tag{2.11}$$

where $g(\eta)$ is an arbitrary function of its argument. The assumption of no rigid zone and the boundary condition (2.2) require $g = 0$. However, a much better solution can be obtained without this assumption. Instead, it is assumed that there is a rigid zone near the axis $x = 0$. The general structure of the kinematically admissible velocity field in the region $0 \le x \le W$ and $0 \le y \le H$ is shown in Fig. 2.3. It is similar to that obtained from the slip-line solution (Fig. 2.2). In general, it is possible to include rigid zone 2 in consideration as well. However, a possible improvement in limit load prediction is negligible. The rigid zone (Fig. 2.3) moves in the direction of the y-axis along with the block of rigid base material. Therefore, the boundary condition (2.2) is satisfied. Also is satisfied the boundary condition (2.4) in the range $0 \le x \le x_b$. The value of x_b will be determined later. The shape of the rigid plastic boundary 0b, which is also a velocity discontinuity curve, should be found from the solution. Let $x = X(y)$ be the equation for this curve. The velocity field is kinematically admissible if and only if this curve contains the origin of the coordinate system. Then, the following condition holds

$$X = 0 \tag{2.12}$$

for $y = 0$.

Let φ be the orientation of the tangent to the velocity discontinuity curve relative to the x-axis. Then, the unit normal vector to this curve can be represented as (Fig. 2.3)

$$\mathbf{n} = -\sin\varphi\mathbf{i} + \cos\varphi\mathbf{j} \tag{2.13}$$

where $\mathbf{i}$ and $\mathbf{j}$ are the base vectors of the Cartesian coordinate system. The velocity vector in the rigid zone is

$$\mathbf{U} = U\mathbf{j}. \tag{2.14}$$

Using Eqs. (2.9) and (2.11) the velocity field in the plastic zone can be written in the form

$$\mathbf{u} = u_x\mathbf{i} + u_y\mathbf{j} = U[-\varsigma/h + g(\eta)]\mathbf{i} + U\eta\mathbf{j}. \tag{2.15}$$

Assume that $\mathbf{U} \equiv \mathbf{u}_1$ and $\mathbf{u} \equiv \mathbf{u}_2$ in Eq. (1.7). Then, it follows from Eqs. (2.13), (2.14) and (2.15) that $[-\varsigma/h + g(\eta)]\sin\varphi + (1-\eta)\cos\varphi = 0$. Since $\tan\varphi = dy/dx$ (Fig. 2.3), this equation can be transformed, with the use of Eq. (2.8), to

$$\frac{d\varsigma}{d\eta} = \frac{\varsigma - hg(\eta)}{1-\eta}. \tag{2.16}$$

This is a linear ordinary differential equation of first order. Therefore, its general solution can be found with no difficulty. The condition (2.12) is equivalent to the condition $\varsigma = 0$ at $\eta = 0$ for Eq. (2.16). The solution to Eq. (2.16) satisfying this condition is

$$\varsigma = \varsigma_{0b}(\eta) = -\frac{h}{(1-\eta)}\int_0^\eta g(\upsilon)d\upsilon \tag{2.17}$$

where υ is a dummy variable. It is worth noting here that the denominator in Eq. (2.17) vanishes at $\eta = 1$. Therefore, the velocity discontinuity curve can have a common point with the line $\eta = 1$ (or $y = H$) if and only if the integral vanishes at $\eta = 1$. Moreover, the right hand side of Eq. (2.17) must tend to a finite limit as $\eta \to 1$. An additional condition for the validity of the subsequent solution is that the x-coordinate of point b (Fig. 2.3) should lie in the range $0 < x_b \le W$. Finally, in order to obtain the structure of the kinematically admissible velocity field shown in Fig. 2.3, it is necessary to impose the following restriction on the function $g(\eta)$

$$\varsigma_b = -h\lim_{\eta\to 1}\left[\frac{1}{(1-\eta)}\int_0^\eta g(\upsilon)d\upsilon\right] = hg(1), \quad 0 < \varsigma_b \le 1. \tag{2.18}$$

Here l'Hospital's rule has been applied and $\varsigma_b = x_b/W$.

The strain rate components and the equivalent strain rate are determined from Eqs. (1.3), (2.8), (2.9), and (2.11) as

$$\zeta_{xx} = \frac{\partial u_x}{\partial x} = -\frac{U}{H}, \quad \zeta_{yy} = \frac{\partial u_y}{\partial y} = \frac{U}{H}, \quad \zeta_{xy} = \frac{1}{2}\left(\frac{\partial u_x}{\partial y} + \frac{\partial u_y}{\partial x}\right) = \frac{Ug'(\eta)}{2H},$$
$$\zeta_{eq} = \frac{1}{\sqrt{3}}\frac{U}{H}\sqrt{4 + [g'(\eta)]^2}$$
(2.19)

where $g'(\eta) \equiv dg(\eta)/d\eta$. The rate of work dissipation in the plastic zone is found with the use of Eq. (2.19) in the form

$$E_V = \sigma_0 \iiint_V \zeta_{eq} dV = \frac{2UBW\sigma_0}{\sqrt{3}} \int_0^1 \int_{\varsigma_{0b}(\eta)}^1 \sqrt{4 + [g'(\eta)]^2} d\varsigma d\eta$$

or, after integration with respect to ς,

$$\frac{E_V}{4UBW\sigma_0} = \frac{1}{2\sqrt{3}} \int_0^1 [1 - \varsigma_{0b}(\eta)]\sqrt{4 + [g'(\eta)]^2} d\eta.$$
(2.20)

There are two velocity discontinuity curves, $0b$ and bc (Fig. 2.3). In the case of line bc, it is assumed that a material layer of infinitesimal thickness sticks at the block of rigid base material according to the boundary condition (2.4). Then, a necessity of the discontinuity follows from Eq. (2.11). Substituting Eqs. (2.14) and (2.15) into Eq. (1.8) gives the amount of velocity jump across the velocity discontinuity curve $0b$ in the form

$$|[u_\tau]|_{0b} = U\sqrt{(1-\eta)^2 + [g(\eta) - \varsigma/h]^2}.$$
(2.21)

Taking into account the boundary condition (2.4) the amount of velocity jump across the velocity discontinuity line bc can be represented as $|u_x|$ at $\eta = 1$ in the range $x_b \le x \le W$ (or $\varsigma_b \le \varsigma \le 1$). Therefore, it follows from Eqs. (2.11) and (2.18) that

$$|[u_\tau]|_{bc} = \frac{U}{h}(\varsigma - \varsigma_b).$$
(2.22)

The rate of work dissipation at the velocity discontinuity curve $0b$ is determined as

$$E_{0b} = \frac{\sigma_0}{\sqrt{3}} \iint_{S_d} |[u_\tau]|_{0b} dS = \frac{2B\sigma_0}{\sqrt{3}} \int_0^H |[u_\tau]|_{0b} \sqrt{1 + \left(\frac{dx}{dy}\right)^2} dy$$

or, with the use of Eq. (2.8),

$$E_{0b} = \frac{\sigma_0}{\sqrt{3}} \iint_{S_d} |[u_\tau]|_{0b} dS = \frac{2BH\sigma_0}{\sqrt{3}} \int_0^1 |[u_\tau]|_{0b} \sqrt{1 + \frac{1}{h^2}\left(\frac{d\varsigma}{d\eta}\right)^2} d\eta.$$
(2.23)

Here the derivative $d\varsigma/d\eta$ should be found at points of the velocity discontinuity curve $0b$ and is therefore given by Eq. (2.16) where ς should be replaced with $\varsigma_{0b}(\eta)$. Substituting Eqs. (2.16) and (2.21) into Eq. (2.23) leads to

$$\frac{E_{0b}}{4UBW\sigma_0} = \frac{h}{2\sqrt{3}} \int_0^1 \frac{(1-\eta)^2 + [g(\eta) - \varsigma_{0b}(\eta)/h]^2}{1-\eta} d\eta. \tag{2.24}$$

Using Eq. (2.8) the rate of work dissipation at the velocity discontinuity line bc is represented in the form

$$E_{bc} = \frac{\sigma_0}{\sqrt{3}} \iint_{S_d} |[u_\tau]|_{bc} dS = \frac{2B\sigma_0}{\sqrt{3}} \int_{x_b}^W |[u_\tau]|_{bc} dx = \frac{2BW\sigma_0}{\sqrt{3}} \int_{\varsigma_b}^1 |[u_\tau]|_{bc} d\varsigma. \tag{2.25}$$

Substituting Eq. (2.22) into Eq. (2.25) and integrating lead to

$$\frac{E_{bc}}{4UBW\sigma_0} = \frac{(1-\varsigma_b)^2}{4\sqrt{3}h}. \tag{2.26}$$

The rate at which forces F do work is

$$\iint_{S_v} (t_i v_i) dS = \frac{1}{2} FU. \tag{2.27}$$

The multiplier 1/2 has appeared here because two equal forces act but one quarter of the specimen is considered. Using Eq. (2.27) the inequality (1.4) for the problem under consideration can be transformed to $F_u U = 2(E_V + E_{0b} + E_{bc})$. Substituting Eqs. (2.20), (2.24) and (2.26) into this equation gives

$$f_u = \frac{F_u}{4BW\sigma_0} = \frac{1}{\sqrt{3}}\left[I_1 + hI_2 + \frac{(1-\varsigma_b)^2}{2h}\right],$$

$$I_1 = \int_0^1 [1 - \varsigma_{0b}(\eta)]\sqrt{4 + [g'(\eta)]^2} d\eta, \tag{2.28}$$

$$I_2 = \int_0^1 \frac{(1-\eta)^2 + [g(\eta) - \varsigma_{0b}(\eta)/h]^2}{1-\eta} d\eta$$

It is now necessary to specify the function $g(\eta)$. It is advantageous to choose this function such that Eq. (1.9) is satisfied in the vicinity of the velocity discontinuity line bc. It is obvious from Eq. (2.19) that the normal strain rates in the Cartesian coordinate system are bounded everywhere. Therefore, it follows from Eq. (1.9) that $|\zeta_{xy}| \to \infty$ or $|\partial u_x/\partial \eta| \to \infty$ as $\eta \to 1$ in the range $\varsigma_b \le \varsigma \le 1$. Moreover, symmetry demands that $g(\eta)$ is an even function of its argument. One of the simplest functions satisfying this symmetry requirement and the inverse square root rule in Eq. (1.9) is

$$g(\eta) = \beta_0 + \beta_1 \sqrt{1 - \eta^2} \qquad (2.29)$$

where β_0 and β_1 are free parameters. Substituting Eq. (2.29) into Eq. (2.18) shows that the limit is finite if and only if $\beta_0 = -\pi\beta_1/4$. Moreover, $\varsigma_b = -h\pi\beta_1/4$. Then, replacing β_0 and β_1 in Eq. (2.29) with ς_b and differentiating give

$$g(\eta) = \frac{\varsigma_b}{h}\left(1 - \frac{4}{\pi}\sqrt{1 - \eta^2}\right), \quad g'(\eta) = \frac{4\varsigma_b}{\pi h}\frac{\eta}{\sqrt{1 - \eta^2}}. \qquad (2.30)$$

Substituting Eq. (2.30) into Eq. (2.17) results in

$$\varsigma_{0b}(\eta) = \frac{\varsigma_b}{\pi(1 - \eta)}\left(2\eta\sqrt{1 - \eta^2} + 2\arcsin\eta - \pi\eta\right). \qquad (2.31)$$

In general, excluding $g(\eta)$, $g'(\eta)$ and $\varsigma_{0b}(\eta)$ on the right hand side of Eq. (2.28) the value of f_u can be evaluated using Eqs. (2.30) and (2.31). However, the integral I_1 involved in Eq. (2.28) is improper since $g'(\eta) \to \infty$ as $\eta \to 1$. Therefore, it is convenient to introduce the new variable ϑ by

$$\eta = \sin\vartheta, \quad d\eta = \cos\vartheta d\vartheta. \qquad (2.32)$$

Then, using Eqs. (2.30) and (2.31) the integral I_1 can be transformed to

$$I_1 = 2\int_0^{\pi/2}\left[1 - \frac{\varsigma_b(\sin 2\vartheta + 2\vartheta - \pi\sin\vartheta)}{\pi(1 - \sin\vartheta)}\right]\sqrt{\cos^2\vartheta + 4\left(\frac{\varsigma_b}{\pi h}\right)^2\sin^2\vartheta}\,d\vartheta. \qquad (2.33)$$

A difficulty with numerical evaluating this integral is that the integrand reduces to the expression $0/0$ at $\vartheta = \pi/2$. In order to facilitate numerical integration, the integrand should be expanded in a series in the vicinity of this point. In particular,

$$\left[1 - \frac{\varsigma_b(\sin 2\vartheta + 2\vartheta - \pi\sin\vartheta)}{\pi(1 - \sin\vartheta)}\right]\sqrt{\cos^2\vartheta + 4\left(\frac{\varsigma_b}{\pi h}\right)^2\sin^2\vartheta}\,d\vartheta$$
$$= \frac{(1 - \varsigma_b)\varsigma_b}{\pi h} + \frac{8\varsigma_b^2}{3\pi^2 h}\left(\frac{\pi}{2} - \vartheta\right) + o\left(\frac{\pi}{2} - \vartheta\right), \quad \vartheta \to \frac{\pi}{2}. \qquad (2.34)$$

Substituting Eq. (2.34) into Eq. (2.33) and integrating analytically over the range $\pi/2 - \delta \le \vartheta \le \pi/2$ lead to

$$I_1 = \frac{2(1 - \varsigma_b)\varsigma_b}{\pi h}\delta + \frac{8\varsigma_b^2}{3\pi^2 h}\delta^2$$
$$+ 2\int_0^{\pi/2-\delta}\left[1 - \frac{\varsigma_b(\sin 2\vartheta + 2\vartheta - \pi\sin\vartheta)}{\pi(1 - \sin\vartheta)}\right]\sqrt{\cos^2\vartheta + 4\left(\frac{\varsigma_b}{\pi h}\right)^2\sin^2\vartheta}\,d\vartheta$$

$$(2.35)$$

where δ is a small number.

Fig. 2.4 Comparison of two solutions for the dimensionless upper bound limit load

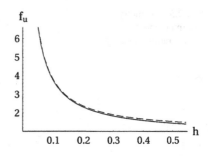

Analogously, it is possible to verify by inspection that the integrand of the integral I_2 involved in Eq. (2.28) reduces to the expression $0/0$ at $\eta = 1$. Using Eqs. (2.30) and (2.31) to exclude $g(\eta)$ and $\varsigma_{0b}(\eta)$ this integral can be rewritten as

$$I_2 = \int\limits_0^1 \frac{\pi^2 h^2 (1 - \eta)^4 + \varsigma_b^2 \left[\dfrac{\left(\pi - 4\sqrt{1 - \eta^2}\right)(1 - \eta) - }{2\eta\sqrt{1 - \eta^2} - 2\arcsin\eta + \pi\eta} \right]^2}{\pi^2 h^2 (1 - \eta)^3} d\eta. \tag{2.36}$$

Expanding the integrand in a series in the vicinity of $\eta = 1$ gives

$$\frac{\pi^2 h^2 (1 - \eta)^4 + \varsigma_b^2 \left[\left(\pi - 4\sqrt{1 - \eta^2}\right)(1 - \eta) - 2\eta\sqrt{1 - \eta^2} - 2\arcsin\eta + \pi\eta \right]^2}{\pi^2 h^2 (1 - \eta)^3}$$

$$= \frac{32\varsigma_b^2}{9\pi^2 h^2} + \left(\frac{16\varsigma_b^2}{5\pi^2 h^2} - 1\right)(\eta - 1) + o(\eta - 1), \quad \eta \to 1. \tag{2.37}$$

Substituting Eq. (2.37) into Eq. (2.36) and integrating lead to

$$I_2 = \frac{1}{2} + \frac{32\varsigma_b^2}{9\pi^2 h^2}\delta - \frac{8\varsigma_b^2}{5\pi^2 h^2}\delta^2$$

$$+ \frac{\varsigma_b^2}{\pi^2 h^2} \int\limits_0^{1-\delta} \frac{\left[\left(\pi - 4\sqrt{1 - \eta^2}\right)(1 - \eta) - 2\eta\sqrt{1 - \eta^2} - 2\arcsin\eta + \pi\eta\right]^2}{(1 - \eta)^3} d\eta.$$

$$\tag{2.38}$$

Integration in Eqs. (2.35) and (2.38) can be performed numerically for any value of ς_b with no difficulty. Then, the right hand side of Eq. (2.28) becomes a function of this parameter. This function should be minimized with respect to ς_b to find the best upper bound based on the kinematically admissible velocity field chosen. This minimization has been carried out numerically assuming that $\delta = 10^{-4}$. Note that the value of ς_b found from this calculation along with f_u is also important because of the restriction $0 < \varsigma_b \leq 1$. It has been found that the latter is not satisfied for $h > 0.54$. The variation of f_u with h determined from Eq. (2.28) after minimization is shown in Fig. 2.4 by the broken line. The solid line

Fig. 2.5 Geometry of the specimen under consideration–notation

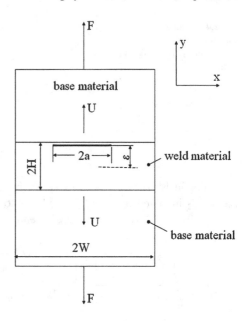

corresponds to the solution given by Eq. (2.5). It is seen from this figure that the limit load found from Eq. (2.28) is just slightly higher than that from the accurate numerical solution. An advantage of the solution (2.28) is that a kinematically admissible velocity field similar to that used to arrive at this solution can be constructed for many other structures with no difficulty whereas numerical solutions are usually time-consuming. Moreover, the slip-line technique used to obtain the solution (2.5) is not applicable to non-planar flow of the material obeying the Mises yield criterion. Nevertheless, since the solution (2.5) is available for the problem under consideration, it will be used in subsequent sections. The solution based on the kinematically admissible velocity field (2.9) and (2.11) has been given to show the main difficulties with using singular kinematically admissible velocity fields for finding the limit load and to demonstrate its accuracy by making comparison with the numerical solution.

2.2 Crack at Some Distance From the Mid-Plane of the Weld

The geometry of the specimen and the direction of the axes of Cartesian coordinates (x, y) are illustrated in Fig. 2.5. The only difference from the previous boundary value problem is that a crack is located at some distance ε from the mid-plane of the weld. Its orientation is orthogonal to the line of action of tensile forces F. It is obvious that the sense of ε is immaterial. Therefore, it is possible to assume that $0 \leq \varepsilon \leq H$. The origin of the coordinate system is located at the intersection of the axes of symmetry of the specimen with no crack. The specimen is symmetric

Fig. 2.6 General structure of the kinematically admissible velocity field

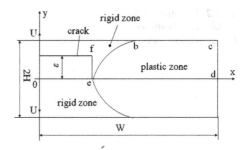

relative to the *y*-axis. Therefore, it is sufficient to find the solution in the domain $x \geq 0$. The center cracked specimen considered in the previous section is obtained at $\varepsilon = 0$. In the case of the interface crack $\varepsilon = H$. A numerical solution for the latter case has been given in Kim and Schwalbe (2001b) where a possible effect of the location of the crack on the limit load is briefly discussed as well.

The general structure of the kinematically admissible velocity field chosen is illustrated in Fig. 2.6 (one half of the weld is shown). The plastic zone is symmetric relative to the *x*-axis. Therefore, it is sufficient to consider its upper part *ebcd*. The rigid zones move along with the blocks of rigid base material along the *y*-axis in the opposite directions. The magnitude of velocity of each zone is *U*. The velocity discontinuity line *ef* separates the two rigid zones. In the plastic zone *ebcd*, the kinematically admissible velocity field can be assumed in the same form as in the zone *0bdc* shown in Fig. 2.2 (the presence of rigid zone 2 is not essential since any rigid zone can be considered as a special case of plastic zones in which $\zeta_{eq} = 0$.) Then, according to Eq. (2.7) the rate of work dissipation in the plastic zone *ebcd* (Fig. 2.3), including the rate of work dissipation at the velocity discontinuity curves *eb* and *bc* as well as any velocity discontinuity curves inside this zone, is given by

$$E_{ebcd} = \frac{UBW\sigma_0}{\sqrt{3}} \left(1 - \frac{a}{W}\right) \left(3 + \frac{W - a}{H}\right). \qquad (2.39)$$

The amount of velocity jump across the velocity discontinuity line *ef* is equal to $|[u_\tau]|_{ef} = 2U$. The area of the corresponding velocity discontinuity surface is $2B\varepsilon$. Therefore, the rate of work dissipation at the velocity discontinuity line *ef* can be found as

$$E_{ef} = \frac{\sigma_0}{\sqrt{3}} \iint_{S_d} |[u_\tau]|_{ef} dS = \frac{2U\sigma_0}{\sqrt{3}} \iint_{S_d} dS = \frac{4UB\varepsilon\sigma_0}{\sqrt{3}}. \qquad (2.40)$$

The rate at which external forces *F* do work is (two equal forces act and one half of the specimen is considered)

$$\iint_{S_v} (t_i v_i) dS = FU. \qquad (2.41)$$

Fig. 2.7 Variation of the
critical crack length with
H/W at different values of ε

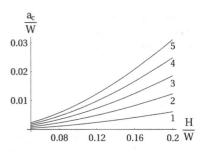

The rate of work dissipation in one half of the specimen is given by $2E_{ebcd} + E_{ef}$.
Therefore, it follows from Eqs. (1.4), (2.39), (2.40), and (2.41) that the upper bound
limit load is

$$f_u = \frac{F_u}{4BW\sigma_0} = \frac{1}{2\sqrt{3}}\left(1 - \frac{a}{W}\right)\left(3 + \frac{W - a}{H}\right) + \frac{\varepsilon}{W\sqrt{3}}. \qquad (2.42)$$

The validity of this solution is restricted by the inequality given after Eq. (2.7).
When this inequality is not satisfied (i.e. in the case of large cracks), the solution is
trivial and is available in the literature (see, for example, Kotousov and Jaffar
2006). The contribution of the last term in Eq. (2.42) can be too large for small
cracks. The kinematically velocity field which has led to the solution (2.5) is also
kinematically admissible for the specimen under consideration. Equating $F_u^{(0)}$ from
Eq. (2.5) and F_u from Eq. (2.42) gives the following equation for the critical value
of $a = a_c$

$$\left(3 + \frac{W}{H}\right) = \left(1 - \frac{a_c}{W}\right)\left(3 + \frac{W - a_c}{H}\right) + \frac{2\varepsilon}{W}. \qquad (2.43)$$

The solution to this quadratic equation is trivial and it is illustrated in Fig. 2.7.
In this figure, curve 1 corresponds to $\varepsilon/H = 0.2$, curve 2 to $\varepsilon/H = 0.4$, curve 3 to
$\varepsilon/H = 0.6$, curve 4 to $\varepsilon/H = 0.8$, and curve 5 to $\varepsilon/H = 1$ (interface crack).
In order to determine the best limit load based on the kinematically admissible
velocity field chosen, Eq. (2.42) should be used for $a \geq a_c$ and Eq. (2.5) for $a \leq a_c$.

2.3 Arbitrary Crack in the Weld

The geometry of the specimen and Cartesian coordinates (x, y) are shown in
Fig. 2.8. The origin of the coordinate system is located at the intersection of the
axes of symmetry of the specimen with no crack. In contrast to the specimens
considered in the previous sections, the crack may have an arbitrary shape, though
some minimal restrictions apply. In particular, it is assumed that the crack is
entirely located within the weld and the shape of the crack does not prevent the
motion of rigid blocks of material below and above the crack in the opposite

Fig. 2.8 Geometry of the
specimen under
consideration–notation

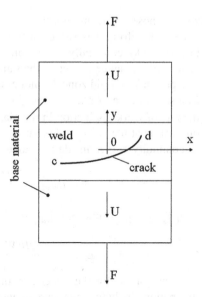

Fig. 2.9 Structure of the
kinematically admissible
velocity field

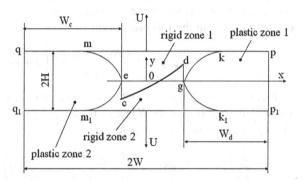

directions along the y-axis. An upper bound solution for the specimen under
consideration has been proposed in Alexandrov (2010).

The crack is specified by the coordinates of its tips. In particular, $x = x_c$ and
$y = y_c$ for tip c, and $x = x_d$ and $y = y_d$ for tip d (Fig. 2.8). By assumption, $x_d \geq 0$
and $x_c \leq 0$. The general structure of the kinematically admissible velocity field
chosen is shown in Fig. 2.9. It consists of two plastic zones and two rigid zones.
Rigid zone 1 whose boundary is *mecdgk* moves along with the block of rigid base
material located above the weld along the positive direction of the y-axis with a
velocity U. Rigid zone 2 whose boundary is m_1ecdgk_1 moves along with the block
of rigid base material located under the weld along the negative direction of the
y-axis with the same velocity. The plastic zones are separated from the rigid zones
by the velocity discontinuity curves me, m_1e, kg, and k_1g. Also, there are four
velocity discontinuity lines at the bi-material interfaces. Those are qm, q_1m_1, kp,
and k_1p_1. Moreover, there are two velocity discontinuity lines separating the rigid

zones. Those are ec and dg. Each of the plastic zones is symmetric relative to the x-axis. Therefore, it is sufficient to consider the upper half of each zone. In particular, the kinematically admissible velocity field in the upper half of the plastic zone $pkgk_1p_1$ can be chosen in the same form as in the zone $0bcd$ shown in Fig. 2.2 (the presence of rigid zone 2 is not essential since any rigid zone can be considered as a special case of plastic zones in which $\zeta_{eq} = 0$). Then, according to Eq. (2.5) the rate of work dissipation in the plastic zone $pkgk_1p_1$, including the rate of work dissipation at the velocity discontinuity curves gk and kp as well as at any velocity discontinuity curves inside this zone, is equal to

$$E_{pkgk_1p_1} = \frac{UBW_d\sigma_0}{\sqrt{3}}\left(3 + \frac{W_d}{H}\right). \tag{2.44}$$

It is seen from Fig. 2.9 that $W_d = W - x_d$. Therefore, Eq. (2.44) becomes

$$E_{pkgk_1p_1} = \frac{UB(W - x_d)\sigma_0}{\sqrt{3}}\left(3 + \frac{W - x_d}{H}\right). \tag{2.45}$$

Analogously, for the upper part of the plastic zone $qmem_1q_1$ the rate of work dissipation, including the rate of work dissipation at the velocity discontinuity curves em and mq as well as at any velocity discontinuity curves inside this zone, can be obtained in the following form

$$E_{qmem_1q_1} = \frac{UB(W + x_c)\sigma_0}{\sqrt{3}}\left(3 + \frac{W + x_c}{H}\right). \tag{2.46}$$

It is worth recalling here that $x_c \leq 0$. The amount of velocity jump across the velocity discontinuity lines df and ec is $\|[u_\tau]\|_{df} = \|[u_\tau]\|_{ec} = 2U$. Therefore, the rates of work dissipation at these lines are

$$E_{df} = \frac{\sigma_0}{\sqrt{3}}\iint\limits_{S_d}\|[u_\tau]\|_{df}dS = \frac{4UB|y_d|\sigma_0}{\sqrt{3}}, \quad E_{ec} = \frac{\sigma_0}{\sqrt{3}}\iint\limits_{S_d}\|[u_\tau]\|_{ec}dS = \frac{4UB|y_c|\sigma_0}{\sqrt{3}}.$$
$$\tag{2.47}$$

The rate at which one external force F does work is given by Eq. (2.41). The rate of total internal work dissipation is $2E_{pkgk_1p_1} + 2E_{qmem_1q_1} + E_{df} + E_{ec}$. Since two identical forces act, it follows from Eqs. (1.4), (2.41), (2.45), (2.46), and (2.47) that

$$\frac{F_u}{4BW\sigma_0} = \frac{(W - x_d)}{4\sqrt{3}W}\left(3 + \frac{W - x_d}{H}\right) + \frac{(W + x_c)}{4\sqrt{3}W}\left(3 + \frac{W + x_c}{H}\right) + \frac{|y_d| + |y_c|}{2\sqrt{3}W}. \tag{2.48}$$

As before, the smallest value between F_u and $F_u^{(0)}$ from Eqs. (2.48) and (2.5), respectively, should be chosen. The solution (2.5) provides a better prediction for small cracks. The solution (2.48) is not valid for large cracks. The restrictions follow from the inequality given after Eq. (2.5) and can be written in the form

$(W - x_d)/H \geq 1$ and $(W + x_c)/H \geq 1$. The solution given in Kotousov and Jaffar (2006) can be adopted when one of these inequalities is not satisfied.

References

S. Alexandrov, A limit load solution for a highly weld strength undermatched tensile panel with an arbitrary crack. Eng. Fract. Mech. **77**, 3368–3371 (2010)

S. Alexandrov, O. Richmond, On estimating the tensile strength of an adhesive plastic layer of arbitrary simply connected contour. Int. J. Solids Struct. **37**, 669–686 (2000)

S. Hao, A. Cornec, K.-H. Schwalbe, Plastic stress-strain fields of a plane strain cracked tensile panel with a mismatched welded joint. Int. J. Solids Struct. **34**, 297–326 (1997)

R. Hill, *The Mathematical Theory of Plasticity* (Clarendon Press, Oxford, 1950)

Y.-J. Kim, K.-H. Schwalbe, Mismatch effect on plastic yield loads in idealised weldments I. Weld centre cracks. Eng. Fract. Mech. **68**, 163–182 (2001a)

Y.-J. Kim, K.-H. Schwalbe, Mismatch effect on plastic yield loads in idealised weldments II. Heat affected zone cracks. Eng. Fract. Mech. **68**, 183–199 (2001b)

A. Kotousov, M.F.M. Jaffar, Collapse load for a crack in a plate with a mismatched welded joint. Eng. Fail. Anal. **13**, 1065–1075 (2006)

L. Prandtl, Anwendungsbeispiele Zu Einem Henckyschen Satz Uber Das Plastische Gleichgewicht. Zeitschr. Angew. Math. Mech. **3**, 401–406 (1923)

Chapter 3
Plane Strain Solutions for Highly Undermatched Scarf Joint Specimens

The specimens considered in this chapter are welded plates with the weld inclined at some angle to the line of action of two tensile forces applied. This angle is denoted by $\pi/2 - \alpha$. By assumption, the constraints imposed require that the rigid blocks of base material move along the line of action of the forces. A crack is entirely located in the weld. Edge cracks are excluded from consideration. The width of the plate is denoted by $2W$, its thickness by $2B$, the thickness of the weld by $2H$, and the length of the crack by $2a$ (except the last solution of this chapter which deals with cracks of arbitrary shape in the plane of flow). The specimens considered in the previous chapter are obtained at $\alpha = 0$. Since plane strain solutions are of concern in this chapter, integration in the thickness direction in volume and surfaces integrals involved in Eq. (1.4) is replaced with the multiplier $2B$ without any further explanation. For the same reason, the term "velocity discontinuity surface" is replaced with the term "velocity discontinuity curve (or line)". The latter refers to curves (lines) in the plane of flow. Base material is supposed to be rigid.

3.1 Specimen with No Crack

The geometry of the specimen and Cartesian coordinates (x, y) are shown in Fig. 3.1. The specimen is loaded by two equal forces F whose magnitude at plastic collapse should be evaluated. The y-axis is perpendicular to the weld. The origin of the coordinate system is located at the intersection of the two axes of symmetry of the specimen wholly made of base material. The blocks of rigid base material move with a velocity U along the line of action of the forces applied. Let u_x be the velocity component in the x-direction and u_y in the y-direction. The velocity boundary conditions are

$$u_y = 0 \tag{3.1}$$

at $y = 0$,

S. Alexandrov, *Upper Bound Limit Load Solutions for Welded Joints with Cracks*, SpringerBriefs in Computational Mechanics, DOI: 10.1007/978-3-642-29234-7_3, © The Author(s) 2012

Fig. 3.1 Geometry of the
specimen under
consideration–notation

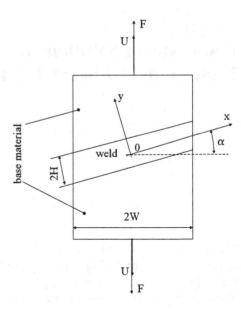

$$u_y = \pm U \cos \alpha \tag{3.2}$$

at $y = \pm H$ and

$$u_x = 0 \tag{3.3}$$

at $y = \pm H$. By analogy to the solution for the specimen with no crack given in the previous chapter, it is reasonable to expect that there is a rigid zone in the vicinity of the y-axis (see Figs. 2.2 and 2.3). A schematic diagram showing the configuration of plastic and rigid zones proposed is presented in Fig. 3.2. It is obvious that plastic zones 1 and 4 are identical in the sense that the rate of internal work dissipation in these zones is the same. Also identical are plastic zones 2 and 3 Therefore, it is sufficient to obtain the solution in the domain $x \geq 0$.

In general, it is possible to generalize the kinematically admissible velocity field given by Eqs. (2.9) and (2.11) to find a solution for the problem under consideration. However, a better result can be obtained by modifying the classical Prandtl's solution for compression of a plastic layer between parallel, rough plates (Prandtl 1923; Hill 1950). The latter has been adopted to arrive at Eq. (2.5). The general structure of the kinematically admissible velocity field is illustrated in Fig. 3.3. Plastic zones 1 and 2 (Fig. 3.2) can now be considered as one plastic zone. It has been shown in Alexandrov and Kontchakova (2007) that the velocity field in this zone generalizing the velocity field proposed in Hill (1950) is

$$\frac{u_x}{U \cos \alpha} = \beta_0 - \frac{x}{H} - 2\sqrt{1 - \frac{y^2}{H^2}} + 2\beta_1 \frac{y}{H}, \quad \frac{u_y}{U \cos \alpha} = \frac{y}{H} \tag{3.4}$$

Fig. 3.2 Configuration of plastic and rigid zones

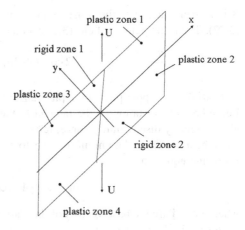

Fig. 3.3 General structure of the kinematically admissible velocity field

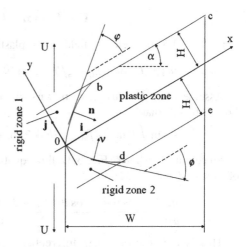

where β_0 and β_1 are arbitrary constants. The solution given in Hill (1950) is obtained as a special case if $\alpha = 0$ and $\beta_1 = 0$. The difference in the sense of the velocity components is immaterial for pressure-independent materials. It is possible to verify by inspection that the velocity field (3.4) satisfies the incompressibility Eq. (1.6) as well as the boundary conditions (3.1) and (3.2). The boundary condition (3.3) is automatically satisfied over the regions of the bi-material interfaces where the weld is rigid. The velocity field is discontinuous over the remainder of the bi-material interfaces. In order to show that the velocity field chosen is kinematically admissible it is just necessary to satisfy the condition that the normal velocity is continuous across the rigid plastic boundaries 0b and 0d (Fig. 3.3). In addition, these curves must cross at the origin of the coordinate system. It is worth noting here that it follows from Eq. (3.4) that $|\xi_{xy}| \to \infty$ as $y \to \pm H$ and Eq. (1.9) is satisfied near the velocity discontinuity lines bc and de.

It is convenient to use the dimensionless quantities introduced in Eqs. (2.8) and (2.32). Then, the velocity field (3.4) becomes

$$\frac{u_x}{U \cos \alpha} = \beta_0 - \frac{\varsigma}{h} - 2 \cos \vartheta + 2\beta_1 \sin \vartheta, \quad \frac{u_y}{U \cos \alpha} = \sin \vartheta. \quad (3.5)$$

Consider the upper part of the plastic zone, $0 \le y \le H$ or $0 \le \vartheta \le \pi/2$ (Fig. 3.3). Let φ be the orientation of the tangent to the rigid plastic boundary $0b$, which is also a velocity discontinuity curve, relative to the x-axis, measured anti-clockwise from the axis. Then, the unit normal vector to this curve can be represented by the following equation

$$\mathbf{n} = \sin \varphi \mathbf{i} - \cos \varphi \mathbf{j} \quad (3.6)$$

where $\mathbf{i}$ and $\mathbf{j}$ are the base vectors of the Cartesian coordinate system. The velocity vector in rigid zone 1 is

$$\mathbf{U} = U \sin \alpha \mathbf{i} + U \cos \alpha \mathbf{j}. \quad (3.7)$$

Using Eq. (3.5) the velocity field in the plastic zone can be written as

$$\mathbf{u} = u_x \mathbf{i} + u_y \mathbf{j} = U \cos \alpha (\beta_0 - \varsigma/h - 2 \cos \vartheta + 2\beta_1 \sin \vartheta)\mathbf{i} + U \cos \alpha \sin \vartheta \mathbf{j}. \quad (3.8)$$

Assume that $\mathbf{U} \equiv \mathbf{u_1}$ and $\mathbf{u} \equiv \mathbf{u_2}$ in Eq. (1.7). Then, it follows from Eqs. (3.6–3.8) that

$$1 - \sin \vartheta + \tan \varphi (\beta_0 - \varsigma/h - 2 \cos \vartheta + 2\beta_1 \sin \vartheta - \tan \alpha) = 0.$$

Since $\tan \varphi = dy/dx$ (Fig. 3.3), this equation can be transformed, with the use of Eqs. (2.8) and (2.32), to

$$\frac{d\varsigma}{d\vartheta} = \frac{\varsigma \cos \vartheta + h \cos \vartheta(-\beta_0 + 2 \cos \vartheta - 2\beta_1 \sin \vartheta + \tan \alpha)}{1 - \sin \vartheta}. \quad (3.9)$$

This is a linear ordinary differential equation of first order. Therefore, its general solution can be found with no difficulty. Since the velocity discontinuity curve $0b$ should contain the origin of the coordinate system, the boundary condition to Eq. (3.9) is $\varsigma = 0$ at $\vartheta = 0$. The solution to Eq. (3.9) satisfying this condition is

$$\varsigma = \varsigma_{0b}(\vartheta) = \frac{h[\vartheta + \sin \vartheta \cos \vartheta + (\tan \alpha - \beta_0) \sin \vartheta - \beta_1 \sin^2 \vartheta]}{(1 - \sin \vartheta)}. \quad (3.10)$$

Since $\vartheta = \pi/2$ at $y = H$, it follows from Eq. (3.10) that the velocity discontinuity curve can have a common point with the line $y = H$ if and only if the right hand side of Eq. (3.10) reduces to the expression $0/0$ at $y = H$. It requires

$$\beta_0 + \beta_1 = \pi/2 + \tan \alpha. \quad (3.11)$$

Consider the lower part of the plastic zone, $-H \le y \le 0$ or $-\pi/2 \le \vartheta \le 0$ (Fig. 3.3). Let ϕ be the orientation of the tangent to the rigid plastic boundary $0d$, which is also a velocity discontinuity curve, relative to the x-axis, measured clockwise from the axis. Then, the unit normal vector to this curve can be represented as

$$\mathbf{v} = \sin\,\phi\mathbf{i} + \cos\,\phi\mathbf{j}. \tag{3.12}$$

The velocity vector in rigid zone 2 is

$$\mathbf{U} = -U\sin\alpha\mathbf{i} - U\cos\alpha\mathbf{j}. \tag{3.13}$$

The velocity field in the plastic zone is given by Eq. (3.8). Assume that $\mathbf{U} \equiv \mathbf{u_1}$ and $\mathbf{u} \equiv \mathbf{u_2}$ in Eq. (1.7). Then, it follows from Eqs. (3.8), (3.12) and (3.13) that

$$1 + \sin\vartheta + \tan\phi(\beta_0 - \varsigma/h - 2\cos\vartheta - 2\beta_1\sin\vartheta + \tan\alpha) = 0.$$

Since $\tan\phi = -dy/dx$ (Fig. 3.3), this equation can be transformed, with the use of Eqs. (2.8) and (2.32), to

$$\frac{d\varsigma}{d\vartheta} = \frac{-\varsigma\cos\vartheta + h\cos\vartheta(\beta_0 - 2\cos\vartheta + 2\beta_1\sin\vartheta + \tan\alpha)}{1 + \sin\vartheta}. \tag{3.14}$$

This is a linear ordinary differential equation of first order. Therefore, its general solution can be found with no difficulty. Since the velocity discontinuity curve $0d$ should contain the origin of the coordinate system, the boundary condition to Eq. (3.14) is $\varsigma = 0$ at $\vartheta = 0$. The solution to Eq. (3.14) satisfying this condition is

$$\varsigma = \varsigma_{0d}(\vartheta) = \frac{h\left[-\vartheta - \sin\vartheta\cos\vartheta + (\tan\alpha + \beta_0)\sin\vartheta + \beta_1\sin^2\vartheta\right]}{(1 + \sin\vartheta)}. \tag{3.15}$$

Since $\vartheta = -\pi/2$ at $y = -H$, it follows from Eq. (3.15) that the velocity discontinuity curve can have a common point with the line $y = -H$ if and only if the right hand side of Eq. (3.15) reduces to the expression $0/0$ at $y = -H$. It requires

$$\beta_0 - \beta_1 = \pi/2 - \tan\alpha. \tag{3.16}$$

Combining Eqs. (3.11) and (3.16) gives

$$\beta_0 = \pi/2, \quad \beta_1 = \tan\alpha. \tag{3.17}$$

Therefore, the velocity field (3.5) becomes

$$\frac{u_x}{U\cos\alpha} = \frac{\pi}{2} - \frac{\varsigma}{h} - 2\cos\vartheta + 2\tan\alpha\sin\vartheta, \quad \frac{u_y}{U\cos\alpha} = \sin\vartheta. \tag{3.18}$$

Using this velocity field along with Eqs. (2.8) and (2.32) the non-zero components of the strain rate tensor can be found as

$$\zeta_{xx} = \frac{\partial u_x}{\partial x} = -\frac{U\cos\alpha}{H}, \quad \zeta_{yy} = \frac{\partial u_y}{\partial y} = \frac{U\cos\alpha}{H},$$

$$\zeta_{xy} = \frac{1}{2}\left(\frac{\partial u_x}{\partial y} + \frac{\partial u_y}{\partial x}\right) = \frac{U\cos\alpha}{H}(\tan\vartheta + \tan\alpha). \tag{3.19}$$

Then, the equivalent strain rate is determined by means of Eq. (1.3) in the form

$$\zeta_{eq} = \frac{2}{\sqrt{3}} \frac{U \cos \alpha}{H} \sqrt{1 + (\tan \vartheta + \tan \alpha)^2}. \tag{3.20}$$

It is evident from Fig. 3.3 that the equation for the line ce is $x \cos \alpha - y \sin \alpha = W$. Using Eqs. (2.8) and (2.32) this equation can be transformed to

$$\varsigma = \varsigma_{ce}(\vartheta) = \frac{1 + h \sin \vartheta \sin \alpha}{\cos \alpha}. \tag{3.21}$$

The rate of work dissipation in the plastic zone follows from Eqs. (2.8), (2.32) and (3.20) in the form

$$E_V = \sigma_0 \iiint_V \zeta_{eq} dV = \frac{4UBW\sigma_0 \cos \alpha}{\sqrt{3}} \int_0^{\pi/2} \int_{\varsigma_{0b}(\vartheta)}^{\varsigma_{ce}(\vartheta)} \cos \vartheta \sqrt{1 + (\tan \vartheta + \tan \alpha)^2} d\varsigma d\vartheta$$

$$+ \frac{4UBW\sigma_0 \cos \alpha}{\sqrt{3}} \int_{-\pi/2}^{0} \int_{\varsigma_{0d}(\vartheta)}^{\varsigma_{ce}(\vartheta)} \cos \vartheta \sqrt{1 + (\tan \vartheta + \tan \alpha)^2} d\varsigma d\vartheta$$

or, after integration with respect to ς,

$$\frac{E_V}{4UBW\sigma_0} = \frac{\cos \alpha}{\sqrt{3}} \int_0^{\pi/2} (\varsigma_{ce}(\vartheta) - \varsigma_{0b}(\vartheta)) \cos \vartheta \sqrt{1 + (\tan \vartheta + \tan \alpha)^2} d\vartheta$$

$$+ \frac{\cos \alpha}{\sqrt{3}} \int_{-\pi/2}^{0} (\varsigma_{ce}(\vartheta) - \varsigma_{0d}(\vartheta)) \cos \vartheta \sqrt{1 + (\tan \vartheta + \tan \alpha)^2} d\vartheta. \tag{3.22}$$

There are four velocity discontinuity curves. Those are $0b$, $0d$, bc and de (Fig. 3.3). Substituting Eq. (3.18) into Eq. (3.8) and, then, Eqs. (3.7) and (3.8) into Eq. (1.8) give the amount of velocity jump across the curve $0b$ in the form

$$|[u_\tau]|_{0b} = U \cos \alpha \sqrt{\left[\tan \alpha(1 - 2 \sin \vartheta) - \frac{\pi}{2} + \frac{\varsigma_{0b}(\vartheta)}{h} + 2 \cos \vartheta\right]^2 + (1 - \sin \vartheta)^2}.$$

$$\tag{3.23}$$

Here ς involved in Eq. (3.18) has been replaced with the function $\varsigma_{0b}(\vartheta)$ which is determined by Eq. (3.10). The rate of work dissipation at this curve is represented as

$$E_{0b} = \frac{\sigma_0}{\sqrt{3}} \iint_{S_d} |[u_\tau]|_{0b} dS = \frac{2B\sigma_0}{\sqrt{3}} \int_0^H |[u_\tau]|_{0b} \sqrt{1 + \left(\frac{dx}{dy}\right)^2} dy$$

or, with the use of Eqs. (2.8) and (2.32),

$$E_{0b} = \frac{2BW\sigma_0}{\sqrt{3}} \int\limits_0^{\pi/2} |[u_\tau]|_{0b} \sqrt{h^2 \cos^2 \vartheta + \left(\frac{d\varsigma}{d\vartheta}\right)^2}\, d\vartheta. \tag{3.24}$$

Here the derivative $d\varsigma/d\vartheta$ should be excluded by means of Eq. (3.9). Then, substituting Eq. (3.23) into Eq. (3.24) results in

$$\frac{E_{0b}}{4UBW\sigma_0} = \frac{h\cos\alpha}{2\sqrt{3}}$$

$$\times \int\limits_0^{\pi/2} \frac{\cos\vartheta}{(1-\sin\vartheta)} \left\{ \left[\begin{array}{c} -\frac{\pi}{2} + \frac{\varsigma_{0b}(\vartheta)}{h} \\ +\tan\alpha\,(1-2\sin\vartheta)+2\cos\vartheta \end{array} \right]^2 +(1-\sin\vartheta)^2 \right\} d\vartheta.$$

$$\tag{3.25}$$

Consider the velocity discontinuity line $0d$ (Fig. 3.3). Substituting Eq. (3.18) into Eq. (3.8) and, then, Eqs. (3.8) and (3.13) into Eq. (1.8) give the amount of velocity jump across the curve $0d$ in the form

$$|[u_\tau]|_{0d}= U\cos\alpha\sqrt{\left[\tan\alpha(1+2\sin\vartheta)+\frac{\pi}{2}-\frac{\varsigma_{0d}(\vartheta)}{h}-2\cos\vartheta\right]^2 +(1+\sin\vartheta)^2}. \tag{3.26}$$

Here ς involved in Eq. (3.18) has been replaced with the function $\varsigma_{0d}(\vartheta)$ which is determined by Eq. (3.15). The rate of work dissipation at this curve is represented as

$$E_{0d} = \frac{\sigma_0}{\sqrt{3}} \iint\limits_{S_d} |[u_\tau]|_{0d}dS = \frac{2B\sigma_0}{\sqrt{3}} \int\limits_{-H}^0 |[u_\tau]|_{0d}\sqrt{1+\left(\frac{dx}{dy}\right)^2}\, dy$$

or, with the use of Eqs. (2.8) and (2.32),

$$E_{0d} = \frac{2BW\sigma_0}{\sqrt{3}} \int\limits_{-\pi/2}^0 |[u_\tau]|_{0d}\sqrt{h^2 \cos^2 \vartheta + \left(\frac{d\varsigma}{d\vartheta}\right)^2}\, d\vartheta. \tag{3.27}$$

Here the derivative $d\varsigma/d\vartheta$ should be excluded by means of Eq. (3.14). Then, substituting Eq. (3.26) into Eq. (3.27) results in

$$\frac{E_{0d}}{4UBW\sigma_0} = \frac{h\cos\alpha}{2\sqrt{3}}$$

$$\times \int_{-\pi/2}^{0} \frac{\cos\vartheta}{(1+\sin\vartheta)} \left\{ \left[\begin{array}{c} \frac{\pi}{2} - \frac{\varsigma_{0d}(\vartheta)}{h} + \\ + \tan\alpha(1+2\sin\vartheta) - 2\cos\vartheta \end{array} \right]^2 + (1+\sin\vartheta)^2 \right\} d\vartheta.$$

$$\text{(3.28)}$$

Since $u_x \neq U\sin\alpha$ at $y = H$ between points b and c (Fig. 3), the interpretation of the boundary condition (3.3) is that a material layer of infinitesimal thickness sticks at the block of rigid base material and there is a velocity jump across this line between points b and c. The amount of this jump is equal to $\|[u_\tau]\|_{bc} = U\sin\alpha - u_x$. Here the velocity component u_x should be determined from Eq. (3.18) at $\vartheta = \pi/2$. Therefore,

$$\|[u_\tau]\|_{bc} = U\cos\alpha\left(\frac{\varsigma}{h} - \frac{\pi}{2} - \tan\alpha\right). \tag{3.29}$$

By definition, $\|[u_\tau]\|_{bc} \geq 0$. Eq. (3.29) has been derived assuming that $U\sin\alpha - u_x \geq 0$. This assumption (or the condition $\|[u_\tau]\|_{bc} \geq 0$ directly) must be verified a posteriori. Using Eq. (2.8) the rate of work dissipation at the velocity discontinuity line bc is represented as

$$E_{bc} = \frac{2B\sigma_0}{\sqrt{3}} \int_{x_b}^{x_c} \|[u_\tau]\|_{bc} dx = \frac{2BW\sigma_0}{\sqrt{3}} \int_{\varsigma_b}^{\varsigma_c} \|[u_\tau]\|_{bc} d\varsigma \tag{3.30}$$

where x_b and x_c are the x-coordinates of points b and c, respectively. Also, $x_b = \varsigma_b W$ and $x_c = \varsigma_c W$. Substituting Eq. (3.29) into Eq. (3.30) and integrating lead to

$$E_{bc} = \frac{BHU\cos\alpha\sigma_0}{\sqrt{3}} \left[\left(\frac{\varsigma_c}{h} - \tan\alpha - \frac{\pi}{2}\right)^2 - \left(\frac{\varsigma_b}{h} - \tan\alpha - \frac{\pi}{2}\right)^2 \right]. \tag{3.31}$$

The value of ς_c follows from Eq. (3.21) at $\vartheta = \pi/2$, $\varsigma_c = \sec\alpha + h\tan\alpha$. The value of ς_b is determined from Eq. (3.10) at $\vartheta = \pi/2$. The right hand side of this equation reduces to the expression $0/0$ at $\vartheta = \pi/2$ when β_0 and β_1 are excluded by means of Eq. (3.17). Applying l'Hospital's rule yields

$$\varsigma_b = \lim_{\vartheta \to \pi/2} \varsigma_{0b}(\vartheta) = h\left(\frac{\pi}{2} + \tan\alpha\right).$$

Substituting the values of ς_c and ς_b found into Eq. (3.31) leads to

$$\frac{E_{bc}}{4UBW\sigma_0} = \frac{h\cos\alpha}{4\sqrt{3}} \left(\frac{1}{h\cos\alpha} - \frac{\pi}{2}\right)^2. \tag{3.32}$$

Also, putting $\varsigma = \varsigma_b$ in Eq. (3.29) and taking into account that the right hand side of this equation is a monotonically increasing function of ς it is possible conclude that the condition $\|[u_\tau]\|_{bc} \geq 0$ is satisfied and, therefore, Eq. (3.29) is valid.

One of the restrictions imposed on the solution based on the velocity field (3.18) is $\varsigma_c \geq \varsigma_b$. Using the found values of these parameters this inequality can be transformed to

$$2 \sec \alpha - h\pi \geq 0. \tag{3.33}$$

As in the case of the line bc, it follows from the boundary condition (3.3) and Eq. (3.18) that the velocity field is discontinuous at $y = -H$ between points d and e. The amount of velocity jump across the velocity discontinuity line de is equal to $|[u_\tau]|_{de} = -U \sin \alpha - u_x$. Here the velocity component u_x should be determined from Eq. (3.18) at $\vartheta = -\pi/2$. Therefore,

$$|[u_\tau]|_{de} = U \cos \alpha \left(\frac{\varsigma}{h} + \tan \alpha - \frac{\pi}{2} \right). \tag{3.34}$$

By definition, $|[u_\tau]|_{de} \geq 0$. Equation (3.34) has been derived assuming that $-U \sin \alpha - u_x \geq 0$. This assumption (or the condition $|[u_\tau]|_{de} \geq 0$ directly) must be verified a posteriori. Using Eq. (2.8) the rate of work dissipation at the velocity discontinuity line de is represented as

$$E_{de} = \frac{2B\sigma_0}{\sqrt{3}} \int\limits_{x_d}^{x_e} |[u_\tau]|_{de} dx = \frac{2BW\sigma_0}{\sqrt{3}} \int\limits_{\varsigma_d}^{\varsigma_e} |[u_\tau]|_{de} d\varsigma \tag{3.35}$$

where x_d and x_e are the x-coordinates of points d and e, respectively. Also, $x_d = \varsigma_d W$ and $x_e = \varsigma_e W$. Substituting Eq. (3.34) into Eq. (3.35) and integrating lead to

$$E_{de} = \frac{BHU \cos \alpha \sigma_0}{\sqrt{3}} \left[\left(\frac{\varsigma_e}{h} + \tan \alpha - \frac{\pi}{2} \right)^2 - \left(\frac{\varsigma_d}{h} + \tan \alpha - \frac{\pi}{2} \right)^2 \right]. \tag{3.36}$$

The value of ς_e follows from Eq. (3.21) at $\vartheta = -\pi/2$, $\varsigma_e = \sec \alpha - h \tan \alpha$. The value of ς_d is determined from Eq. (3.15) at $\vartheta = -\pi/2$. The right hand side of this equation reduces to the expression $0/0$ at $\vartheta = -\pi/2$ when β_0 and β_1 are excluded by means of Eq. (3.17). Applying l'Hospital's rule yields

$$\varsigma_d = \lim_{\vartheta \to -\pi/2} \varsigma_{0d}(\vartheta) = h \left(\frac{\pi}{2} - \tan \alpha \right).$$

Substituting the values of ς_e and ς_d found into Eq. (3.36) leads to

$$\frac{E_{de}}{4UBW\sigma_0} = \frac{h \cos \alpha}{4\sqrt{3}} \left(\frac{1}{h \cos \alpha} - \frac{\pi}{2} \right)^2. \tag{3.37}$$

Also, putting $\varsigma = \varsigma_d$ in Eq. (3.34) and taking into account that the right hand side of this equation is a monotonically increasing function of ς, it is possible to conclude that the condition $|[u_\tau]|_{de} \geq 0$ is satisfied.

One of the restrictions imposed on the solution based on the velocity field (3.18) is $\varsigma_e \geq \varsigma_d$. Using the found values of these parameters this inequality can be reduced to Eq. (3.33).

The rate at which two external forces F do work is (for one half of the specimen)

$$\iint\limits_{S_v} (t_i v_i)\, dS = FU. \tag{3.38}$$

Using Eq. (3.38) the inequality (1.4) for the problem under consideration can be written in the form

$$f_u = \frac{F_u}{4BW\sigma_0} = \frac{E_V + E_{0b} + E_{0d} + E_{bc} + E_{de}}{4UBW\sigma_0}. \tag{3.39}$$

The right hand side of this equation can be found numerically by means of Eqs. (3.22), (3.25), (3.28), (3.32) and (3.37) with the use of Eqs. (3.10), (3.15), (3.17) and (3.21). The final expression contains no free parameters. Therefore, the upper bound limit load is determined by numerical integration with no minimization. It is possible to verify by inspection that the integrands in Eqs. (3.25) and (3.28) reduce to the expression $0/0$ at $\vartheta = \pi/2$ and $\vartheta = -\pi/2$, respectively. Passing to the limit shows that both vanish at the corresponding points. This should be taken into account to facilitate numerical integration. The variation of the dimensionless upper bound limit load with h at different values of α is illustrated in Fig. 3.4 where curve 1 corresponds to $\alpha = 0$, curve 2 to $\alpha = \pi/6$, curve 3 to $\alpha = \pi/4$, and curve 4 to $\alpha = \pi/3$. The restriction (3.33) has been checked in course of calculation. The difference between the solution for $\alpha = 0$ (curve 1) and the solution (2.5) is very small. This indicates that the accuracy of the solution found is rather high.

3.2 Crack at the Mid-Plane of the Weld

The geometry of the specimen is illustrated in Fig. 3.5. The only difference from the previous boundary value problem is that a crack is located at the mid-plane of the weld. Its center lies on the vertical axis of symmetry of the specimen with no crack wholly made of based material. A schematic diagram showing the configuration of plastic and rigid zones proposed is shown in Fig. 3.6. Plastic zones 1 and 2 are identical in the sense that the rate of internal work dissipation in these zones is the same. Comparing the configurations of the plastic zone in Fig. 3.3 and plastic zone 1 in Fig. 3.6 it is possible to conclude that Eq. (3.39) is valid for the specimen under consideration if W is replaced with W_1. It is seen from Fig. 3.6 that

$$W_1 = W - a \cos \alpha. \tag{3.40}$$

Fig. 3.4 Variation of the dimensionless upper bound limit load with h at different α-values

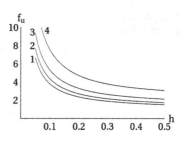

Fig. 3.5 Geometry of the specimen under consideration–notation

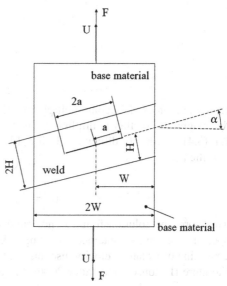

Note that W is involved in the definition (2.8) for h and, moreover, the right hand side of Eq. (3.39) depends on h. Therefore, using Eqs. (3.39), (3.40) for the specimen under consideration can be transformed to

$$f_u = \frac{F_u}{4BW\sigma_0} = \left(1 - \frac{a}{W}\cos\alpha\right)\left(\frac{E_V + E_{0b} + E_{0d} + E_{bc} + E_{de}}{4UBW_1\sigma_0}\right). \tag{3.41}$$

Here $E_V/(4UBW_1\sigma_0)$, $E_{0b}/(4UBW_1\sigma_0)$, $E_{0d}/(4UBW_1\sigma_0)$, $E_{bc}/(4UBW_1\sigma_0)$ and $E_{de}/(4UBW_1\sigma_0)$ should be found by means of the right hand side of Eqs. (3.22), (3.25), (3.28), (3.32) and (3.37), respectively, where h should be replaced with h_1 whose value is determined by

$$h_1 = \frac{H}{W_1} = h\left(1 - \frac{a}{W}\cos\alpha\right)^{-1}. \tag{3.42}$$

Here Eq. (3.40) has been taken into account. Moreover, Eqs. (3.10), (3.15), (3.17) and (3.21) should be used to exclude $\varsigma_{0b}(\vartheta)$, $\varsigma_{0d}(\vartheta)$, $\varsigma_{ce}(\vartheta)$, β_0 and β_1 where it is necessary. Replacing h with h_1 in Fig. 3.4 provides a geometric

Fig. 3.6 Configuration of plastic and rigid zones

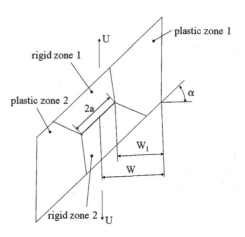

representation of the second multiplier on the right hand side of Eq. (3.41). Having the magnitude of this multiplier the upper bound limit load can be found from Eq. (3.41) with no difficulty. Using Eq. (3.42) the inequality (3.33) can be rewritten as

$$\frac{a}{W} \le \sec \alpha - \frac{\pi h}{2}. \qquad (3.43)$$

Therefore, the solution found is not valid for large cracks. The solution for this special case can be obtained by using a kinematically admissible velocity field consisting of isolated velocity discontinuity lines. It is trivial and is available in the literature (Kotousov and Jaffar 2006; Alexandrov and Kontchakova 2007).

3.3 Crack at Some Distance from the Mid-Plane of the Weld

The geometry of the specimen is shown in Fig. 3.7. The only difference from the previous boundary value problem is that the crack is now located at some distance ε from the mid-plane of the weld. The center of the crack lies on the vertical axis of symmetry of the specimen with no crack wholly made of base material. It is possible to assume, with no loss of generality, that the crack is located above the mid-plane of the weld. Then, $0 \le \varepsilon \le H$. As in the previous section, it is sufficient to consider one half of the specimen. The general structure of the kinematically admissible velocity field chosen is shown in Fig. 3.8. It consists of one plastic zone, two rigid zones and five velocity discontinuity curves. The latter are $0b$, $0d$, bc, de, and $0f$.

Let E_t be the rate of total internal work dissipation in the plastic zone and at the velocity discontinuity curves $0b$, $0d$, bc, and de. The rate at which two forces F do work is given by Eq. (3.38). Therefore, the inequality (1.4) for the problem under consideration can be rewritten as

Fig. 3.7 Geometry of the specimen under consideration–notation

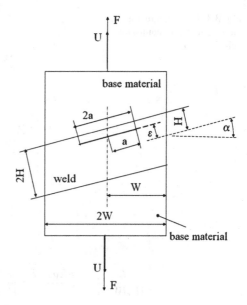

$$f_u = \frac{F_u}{4BW\sigma_0} = \frac{E_t + E_{0f}}{4UBW\sigma_0} \tag{3.44}$$

where E_{0f} is the rate of work dissipation at the velocity discontinuity line $0f$. This line separates two rigid zones moving in the opposite directions. The velocity of each zone is U. Therefore, the amount of velocity jump across this line is equal to $\|[u_\tau]\|_{0f} = 2U$. Then, the rate of work dissipation at this velocity discontinuity line is given by

$$E_{0f} = \frac{\sigma_0}{\sqrt{3}} \iint\limits_{S_d} \|[u_\tau]\|_{0f} dS = \frac{4UB\varepsilon\sigma_0}{\sqrt{3}}. \tag{3.45}$$

It is convenient to consider the following identity

$$\frac{E_t}{4UBW\sigma_0} = \frac{W_1}{W}\left(\frac{E_t}{4UBW_1\sigma_0}\right) = \left(1 - \frac{a}{W}\cos\alpha\right)\left(\frac{E_t}{4UBW_1\sigma_0}\right) \tag{3.46}$$

where W_1 is determined by the location of the crack tip as shown in Fig. 3.8. Analytically, it is given by Eq. (3.40). Substituting Eqs. (3.45) and (3.46) into Eq. (3.44) gives

$$f_u = \left(1 - \frac{a}{W}\cos\alpha\right)\left(\frac{E_t}{4UBW_1\sigma_0}\right) + \frac{\varepsilon}{\sqrt{3}W}. \tag{3.47}$$

Comparing the configurations of the plastic zones in Figs. 3.3 and 3.8 and recalling the definition for E_t it is possible to conclude that

Fig. 3.8 General structure of the kinematically admissible velocity field

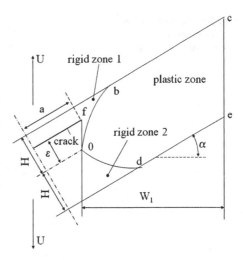

$$\frac{E_t}{4UBW_1\sigma_0} = \left(\frac{E_V + E_{0b} + E_{0d} + E_{bc} + E_{de}}{4UBW_1\sigma_0}\right). \tag{3.48}$$

Here $E_V/(4UBW_1\sigma_0)$, $E_{0b}/(4UBW_1\sigma_0)$, $E_{0d}/(4UBW_1\sigma_0)$, $E_{bc}/(4UBW_1\sigma_0)$, and $E_{de}/(4UBW_1\sigma_0)$ should be found by means of the right hand side of Eqs. (3.22), (3.25), (3.28), (3.32) and (3.37), respectively, where h should be replaced with h_1 given by Eq. (3.42). Substituting the result of this calculation into Eq. (3.47) the upper bound limit load for the structure under consideration can be found. Replacing h with h_1 in Fig. 3.4 provides a geometric representation of the left hand side of Eq. (3.48).

The solution is not valid for large cracks when the inequality (3.33) is not satisfied. The solution for this special case can be obtained by using a kinematically admissible velocity field consisting of isolated velocity discontinuity lines. Similar solutions are available in the literature (Kotousov and Jaffar 2006; Alexandrov and Kontchakova 2007).

For short cracks, the solutions given by Eqs. (3.39) and (3.47) should be compared and the smaller value of f_u should be chosen.

Another restriction on the solution is that the velocity discontinuity curve $0b$ must be on the right to the velocity discontinuity line $0f$ (Fig. 3.8). This restriction is satisfied if $\tan\varphi \leq \tan(\pi/2 - \alpha)$ at $x = 0$ (angle φ has been introduced in Fig. 3.3). Using Eqs. (2.8), (2.32), (3.9) and (3.17) it is possible to show that this inequality is never violated.

3.4 Arbitrary Crack in the Weld

The geometry of the specimen is shown in Fig. 3.9. A crack of quite arbitrary shape is entirely located in the weld. It is convenient to introduce two Cartesian coordinate systems (x, y) and (x_1, y_1). The origin of both coordinate systems coincides with the

Fig. 3.9 Geometry of the
specimen under
consideration–notation

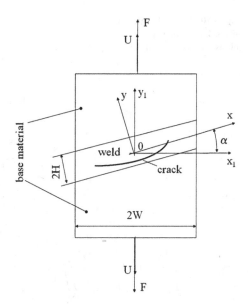

intersection of two axes of symmetry of the specimen with no crack wholly made of
base material. The y-axis is orthogonal to the weld and the bi-material interfaces are
determined by the equations $y = \pm H$. The general structure of the kinemtically
admissible velocity field proposed is illustrated in Fig. 3.10 (base material is not
shown). It consists of two plastic zones, two rigid zones and velocity discontinuity
curves $k_1 b_1$, $b_1 c_1$, $k_1 d_1$, $d_1 e_1$, $k_1 f_1$, $k_2 b_2$, $b_2 c_2$, $k_2 d_2$, $d_2 e_2$ and $k_2 f_2$. The main
restriction on the shape of the crack is that the rigid zones may move along the y_1-axis
(Fig. 3.9) in the opposite directions. Under this restriction, it is sufficient to specify
the coordinates of the crack tips to solve the problem. Let (x_{f1}, y_{f1}) be the coordinates of point f_1 and (x_{f2}, y_{f2}) be the coordinates of point f_2 in the (x, y)—coordinate
system. An additional restriction on the shape of the crack accepted here is that
$x_{f1} \geq 0$ and $x_{f2} \leq 0$. An upper bound solution to the problem formulated has been
proposed by Alexandrov (2011). This solution is given below.

Consider the domain $x \geq 0$. Let E_1 be the rate of total internal work dissipation
in the plastic zone and at the velocity discontinuity curves $k_1 b_1$, $b_1 c_1$,
$k_1 d_1$ and $d_1 e_1$. Then, the rate of total internal work dissipation in the domain under
consideration is $E_1 + E_{k_1 f_1}$ where $E_{k_1 f_1}$ is the rate of work dissipation at the velocity
discontinuity line $k_1 f_1$.

Analogously, let E_2 be the rate of total internal work dissipation in plastic zone
2 and at the velocity discontinuity curves $k_2 b_2$, $b_2 c_2$, $k_2 d_2$. and $d_2 e_2$. (Fig. 3.10).
Then, the rate of total internal work dissipation in the domain $x \leq 0$ is $E_2 + E_{k_2 f_2}$
where $E_{k_2 f_2}$ is the rate of work dissipation at the velocity discontinuity line $k_2 f_2$.
Since the entire specimen is considered, Eq. (3.38) for the rate at which external
forces F do work should be replaced with

Fig. 3.10 General structure
of the kinematically
admissible velocity field

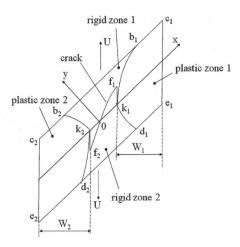

$$\iint\limits_{S_v} (t_i v_i)\, dS = 2FU. \tag{3.49}$$

Therefore, the inequality (1.4) for the problem under consideration becomes

$$f_u = \frac{F_u}{4BW\sigma_0} = \frac{E_1 + E_2 + E_{k_1 f_1} + E_{k_2 f_2}}{8UBW\sigma_0}. \tag{3.50}$$

The velocity discontinuity line $k_1 f_1$ separates two rigid zones moving in the opposite directions. The velocity of each zone is U. Therefore, the amount of velocity jump across the velocity discontinuity line $k_1 f_1$ is $|[u_\tau]|_{k_1 f_1} = 2U$. Then, the rate of work dissipation at this line is given by

$$E_{k_1 f_1} = \frac{\sigma_0}{\sqrt{3}} \iint\limits_{S_d} |[u_\tau]|_{k_1 f_1}\, dS = \frac{4UBL_1\sigma_0}{\sqrt{3}} \tag{3.51}$$

where L_1 is the length of the velocity discontinuity line $k_1 f_1$ (Fig. 3.10). This line is parallel to the y_1-axis and the y-coordinate of point k_1 is $y_{k1} = 0$ (Fig. 3.9). Therefore, using the transformation equations

$$x_1 = x \cos \alpha - y \sin \alpha, \quad y_1 = x \sin \alpha + y \cos \alpha \tag{3.52}$$

the x-coordinate of point k_1, x_{k1}, and L_1 are determined as

$$x_{k1} = x_{f1} - y_{f1} \tan \alpha, \quad L_1 = |y_{f1}| / \cos \alpha \tag{3.53}$$

Analogously, the rate of work dissipation at the velocity discontinuity line $k_2 f_2$ is given by

$$E_{k_2 f_2} = \frac{\sigma_0}{\sqrt{3}} \iint\limits_{S_d} |[u_\tau]|_{k_2 f_2}\, dS = \frac{4UBL_2\sigma_0}{\sqrt{3}} \tag{3.54}$$

where L_2 is the length of line $k_2 f_2$ (Fig. 3.10). This line is parallel to the y_1-axis (Fig. 3.9). Therefore, using the transformation Eq. (3.52) and taking into account that the y-coordinate of point k_2 is $y_{k2} = 0$ it is possible to find that

$$x_{k2} = x_{f2} - y_{f2} \tan \alpha, \quad L_2 = |y_{f2}|/\cos \alpha. \tag{3.55}$$

It is convenient to consider the following identities

$$\frac{E_1}{8UBW\sigma_0} = \frac{W_1}{W} \left(\frac{E_1}{8UBW_1\sigma_0} \right), \quad \frac{E_2}{8UBW\sigma_0} = \frac{W_2}{W} \left(\frac{E_2}{8UBW_2\sigma_0} \right) \tag{3.56}$$

where W_1 and W_2 are determined by the location of the crack tips as shown in Fig. 3.10. In particular, it follows from Eqs. (3.53) and (3.55) that

$$W_1 = W - x_{k1} \cos \alpha = W - x_{f1} \cos \alpha + y_{f1} \sin \alpha,$$
$$W_2 = W + x_{k2} \cos \alpha = W + x_{f2} \cos \alpha - y_{f2} \sin \alpha. \tag{3.57}$$

Substituting Eq. (3.57) into Eq. (3.56) and, then, Eqs. (3.51), (3.54) and (3.56) into Eq. (3.50) with the use of Eqs. (3.53) and (3.55) give

$$f_u = \frac{1}{2} \left(1 - \frac{x_{f1}}{W} \cos \alpha + \frac{y_{f1}}{W} \sin \alpha \right) \left(\frac{E_1}{4UBW_1\sigma_0} \right) + \frac{|y_{f1}/W|}{2\sqrt{3} \cos \alpha}$$
$$+ \frac{1}{2} \left(1 + \frac{x_{f2}}{W} \cos \alpha - \frac{y_{f2}}{W} \sin \alpha \right) \left(\frac{E_2}{4UBW_2\sigma_0} \right) + \frac{|y_{f2}/W|}{2\sqrt{3} \cos \alpha}. \tag{3.58}$$

Comparing the configurations of the plastic zones in Figs. 3.3 and 3.10 and recalling the definition for E_1 and E_2 it is possible to conclude that

$$\frac{E_1}{4UBW_1\sigma_0} = \left(\frac{E_V + E_{0b} + E_{0d} + E_{bc} + E_{de}}{4UBW_1\sigma_0} \right),$$
$$\frac{E_2}{4UBW_2\sigma_0} = \left(\frac{E_V + E_{0b} + E_{0d} + E_{bc} + E_{de}}{4UBW_2\sigma_0} \right). \tag{3.59}$$

Here $E_V/(4UBW_1\sigma_0)$, $E_{0b}/(4UBW_1\sigma_0)$, $E_{0d}/(4UBW_1\sigma_0)$, $E_{bc}/(4UBW_1\sigma_0)$, $E_{de}/(4UBW_1\sigma_0)$, $E_V/(4UBW_2\sigma_0)$, $E_{0b}/(4UBW_2\sigma_0)$, $E_{0d}/(4UBW_2\sigma_0)$, $E_{bc}/(4UBW_2\sigma_0)$, and $E_{de}/(4UBW_2\sigma_0)$ should be found by means of the right hand side of Eqs. (3.22), (3.25), (3.28), (3.32) and (3.37), respectively, where h should be replaced with $h_1 = H/W_1$ or $h_2 = H/W_2$. Using Eq. (3.57) these parameters can be determined for any given crack with no difficulty. Since the coordinates of the crack tips are supposed to be known, substituting E_1 and E_2 found using Eq. (3.59) into Eq. (3.58) results in the dimensionless upper bound limit load for the structure under consideration.

It has been assumed that $x_{f1} \geq 0$ and $x_{f2} \leq 0$. Nevertheless, the solution found is formally valid even if $x_{f1} < 0$ (or $x_{f2} > 0$). However, the larger $|x_{f1}|$ (or x_{f2}) in the case of $x_{f1} < 0$ (or $x_{f2} > 0$), the less accurate solution is. Therefore, it is recommended to use the solution for specimens with $x_{f1} \geq 0$ and $x_{f2} \leq 0$.

The restrictions for short and large cracks are the same as those formulated in the previous section. In particular, in the case of large cracks a kinematically admissible velocity field consisting of isolated velocity discontinuity lines provides an accurate solution. The technique here is similar to that adopted in Kotousov and Jaffar (2006). For specimens with short cracks, f_u from Eq. (3.58) should be compared to f_u from Eq. (3.39) and the smaller value should be adopted as the limit load.

References

S. Alexandrov, A limit load solution for a highly undermatched scarf-joint specimen with an arbitrary crack. Fatigue Fract. Eng. Mater. Struct. **34**, 619–623 (2011)

S. Alexandrov, N. Kontchakova, Upper bound estimate of tension load in tension of welded and brazed plates having an oblique fillet weld seam containing a crack. J. Mach. Manuf. Reliab. **36**, 50–56 (2007)

R. Hill, *The Mathematical Theory of Plasticity* (Clarendon Press, Oxford, 1950)

A. Kotousov, M.F.M. Jaffar, Collapse load for a crack in a plate with a mismatched welded joint. Eng. Fail. Anal. **13**, 1065–1075 (2006)

L. Prandtl, Anwendungsbeispiele Zu Einem Henckyschen Satz Uber Das Plastische Gleichgewicht. Zeitschr. Angew. Math. Mech. **3**, 401–406 (1923)

Chapter 4
Axisymmetric Solutions for Highly Undermatched Tensile Specimens

The specimens considered in this chapter are axisymmetric welded solid or hollow circular cylinders with the weld orientation orthogonal to the axis of symmetry which is also the line of action of forces applied. An axisymmetric crack is entirely located in the weld. The outer radius of the cylinder is denoted by R, the inner radius (when it is applicable) by R_0, and the thickness of the weld by $2H$. Also, u_r stands for the radial velocity and u_z for the axial velocity in a cylindrical coordinate system (r, θ, z). The solutions are independent of θ. Therefore, velocity discontinuity surfaces are referred to as velocity discontinuity curves (or lines) which are in fact the intersections of the velocity discontinuity surfaces and a plane $\theta = \text{constant}$. Moreover, integration with respect to θ in volume and surface integrals involved in Eq. (1.4) is automatically replaced by the multiplier 2π without any further explanation. Base material is supposed to be rigid.

4.1 Solid Cylinder with No Crack

The geometry of the specimen and the cylindrical coordinate system are illustrated in Fig. 4.1. The specimen is loaded by two equal forces F whose magnitude at plastic collapse should be evaluated. The blocks of rigid base material move with a velocity U along the z-axis in the opposite directions. The problem is symmetric relative to the plane $z = 0$. Therefore, it is sufficient to find the solution in the domain $z \geq 0$. The velocity boundary conditions are

$$u_z = 0 \tag{4.1}$$

at $z = 0$,

$$u_z = U \tag{4.2}$$

S. Alexandrov, *Upper Bound Limit Load Solutions for Welded Joints with Cracks*, SpringerBriefs in Computational Mechanics, DOI: 10.1007/978-3-642-29234-7_4, © The Author(s) 2012

Fig. 4.1 Geometry of the
specimen under
consideration–notation

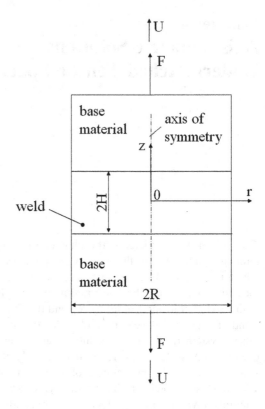

at $z = H$, and

$$u_r = 0 \qquad (4.3)$$

at $r = 0$ and $z = H$. A rigid zone should appear in the vicinity of the axis of symmetry in the exact solution. Therefore, it is reasonable to assume the existence of such a zone in the present upper bound solution. A schematic diagram showing the general structure of the proposed kinematically admissible velocity field in one half of the cross-section of the weld by a plane $\theta = $ constant is presented in Fig. 4.2. The rigid zone moves along the z-axis along with the block of rigid base material. Therefore, the boundary condition (4.3) at $r = 0$ is automatically satisfied. Also is satisfied the boundary condition (4.3) at $z = H$ but only in the domain $0 \leq r \leq r_b$. The value of r_b will be determined later. The shape of the velocity discontinuity curve $0b$ should be found from the solution. It is convenient to introduce the following dimensionless quantities

$$\frac{z}{H} = \eta, \quad \frac{r}{R} = \rho, \quad \frac{H}{R} = h. \qquad (4.4)$$

By analogy to the corresponding plane strain solution given in Chap. 2 the axial velocity can be assumed in the form

Fig. 4.2 General structure of
the kinematically admissible
velocity field

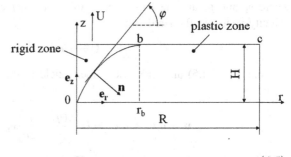

$$u_z = U\eta. \tag{4.5}$$

It is evident that the boundary conditions (4.1) and (4.2) are automatically satisfied. The equation of incompressibility (1.6) in the case under consideration reduces to

$$\frac{\partial u_r}{\partial r} + \frac{u_r}{r} + \frac{\partial u_z}{\partial z} = 0. \tag{4.6}$$

Using Eqs. (4.4) and (4.5) Eq. (4.6) can be transformed to

$$h\frac{\partial(\rho u_r)}{\rho\partial\rho} = -U. \tag{4.7}$$

This equation can be immediately integrated to give

$$\frac{u_r}{U} = \frac{g(\eta)}{\rho} - \frac{\rho}{2h} \tag{4.8}$$

where $g(\eta)$ is an arbitrary function of η. The non-zero components of the strain rate tensor and the equivalent strain rate are determined from Eqs. (1.3), (4.4), (4.5) and (4.8) as

$$\zeta_{rr} = \frac{\partial u_r}{\partial r} = -\frac{U}{R}\left[\frac{g(\eta)}{\rho^2} + \frac{1}{2h}\right], \quad \zeta_{\theta\theta} = \frac{u_r}{r} = \frac{U}{R}\left[\frac{g(\eta)}{\rho^2} - \frac{1}{2h}\right], \quad \zeta_{zz} = \frac{\partial u_z}{\partial z} = \frac{U}{H},$$

$$\zeta_{rz} = \frac{1}{2}\left(\frac{\partial u_r}{\partial z} + \frac{\partial u_z}{\partial r}\right) = \frac{Ug'(\eta)}{2H\rho}, \quad \zeta_{eq} = \frac{U}{H}\sqrt{\frac{4h^2[g(\eta)]^2}{3\rho^4} + \frac{[g'(\eta)]^2}{3\rho^2} + 1}. \tag{4.9}$$

Here and in what follows $g'(\eta) \equiv dg/d\eta$.

Let $\rho = \rho_{0b}(\eta)$ be the equation for the velocity discontinuity curve $0b$ and φ be the orientation of the tangent to this curve relative to the r-axis, measured anti-clockwise from the axis (Fig. 4.2). Then, the unit normal vector to this curve can be represented by the following equation

$$\mathbf{n} = \sin\varphi\mathbf{e_r} - \cos\varphi\mathbf{e_z} \tag{4.10}$$

where $\mathbf{e_r}$ and $\mathbf{e_z}$ are the base vectors of the cylindrical coordinate system. The velocity vector in the rigid zone is

$$\mathbf{U} = U\mathbf{e_z}. \tag{4.11}$$

Using Eqs. (4.5) and (4.8) the velocity field in the plastic zone can be represented as

$$\mathbf{u} = u_r\mathbf{e_r} + u_z\mathbf{e_z} = U\left[\frac{g(\eta)}{\rho} - \frac{\rho}{2h}\right]\mathbf{e_r} + U\eta\mathbf{e_z}. \tag{4.12}$$

Assume that $\mathbf{U} \equiv \mathbf{u_1}$ and $\mathbf{u} \equiv \mathbf{u_2}$ in Eq. (1.7). Then, it is possible to find from Eqs. (4.10), (4.11) and (4.12) that

$$(1 - \eta)\cos\varphi + \left[\frac{g(\eta)}{\rho} - \frac{\rho}{2h}\right]\sin\varphi = 0. \tag{4.13}$$

Since $\tan\varphi = dz/dr$ (Fig. 4.2), Eq. (4.13) becomes

$$(1 - \eta)\frac{d\rho}{d\eta} + \frac{hg(\eta)}{\rho} - \frac{\rho}{2} = 0. \tag{4.14}$$

Here Eq. (4.4) has been taken into account. The velocity field proposed is kinematically admissible if and only if the velocity discontinuity curve $0b$ contains the origin of the coordinate system (Fig. 4.2). Therefore, the boundary condition to Eq. (4.14) is $\rho = 0$ for $\eta = 0$. The solution of Eq. (4.14) satisfying this condition is

$$\rho = \rho_{0b}(\eta) = \sqrt{-\frac{2h}{(1 - \eta)}\int_0^\eta g(\upsilon)d\upsilon} \tag{4.15}$$

where υ is a dummy variable. It is worth noting here that the denominator in Eq. (4.15) vanishes at $\eta = 1$. Therefore, the velocity discontinuity curve $0b$ has a common point with the line $z = H$ if and only if the integral vanishes at $\eta = 1$. Moreover, the right hand side of Eq. (4.15) must tend to a finite limit as $\eta \rightarrow 1$. An additional condition is that the r-coordinate of point b (Fig. 4.2) should be in the range $0 < r_b \leq R$ (or $0 < \rho_b \leq 1$ where $R\rho_b = r_b$). In summary, in order to obtain the structure of the kinematically admissible velocity field shown in Fig. 4.2, it is necessary to impose the following restriction on the function $g(\eta)$

$$\rho_b = \lim_{\eta \rightarrow 1}\sqrt{-\frac{2h}{(1 - \eta)}\int_0^\eta g(\upsilon)d\upsilon} = \sqrt{2hg(1)}, \quad 0 < \rho_b \leq 1. \tag{4.16}$$

Here l'Hospital's rule has been applied.

Using Eqs. (4.4) and (4.9) the rate of work dissipation in the plastic zone is determined as

$$E_V = \sigma_0 \iiint\limits_V \zeta_{eq} dV = 2\pi U R^2 \sigma_0 I_2,$$

$$I_2 = \int\limits_0^1 \int\limits_{\rho_{0b}(\eta)}^1 \sqrt{\frac{4h^2[g(\eta)]^2}{3\rho^4} + \frac{[g'(\eta)]^2}{3\rho^2} + 1}\, \rho d\rho d\eta. \tag{4.17}$$

There are two velocity discontinuity curves, $0b$ and bc (Fig. 4.2). In the case of line bc, it is assumed that a material layer of infinitesimal thickness sticks at the block of rigid base material according to the boundary condition (4.3) at $z = H$. Since u_r given in Eq. (4.8) does not vanish at $z = H$, this discontinuity occurs. Substituting Eqs. (4.11) and (4.12) into Eq. (1.8) results in the amount of velocity jump across the velocity discontinuity curve $0b$ in the form

$$|[u_\tau]|_{0b} = U\sqrt{(1-\eta)^2 + \left[\frac{g(\eta)}{\rho} - \frac{\rho}{2h}\right]^2}. \tag{4.18}$$

As follows from the boundary condition (4.3) at $z = H$, the amount of velocity jump across the velocity discontinuity line bc is equal to $|u_r|$ at $\eta = 1$ where u_r should be found from Eq. (4.8). Therefore,

$$|[u_\tau]|_{bc} = \frac{U}{\rho}\left|\frac{\rho^2}{2h} - g(1)\right|. \tag{4.19}$$

It is evident from Eqs. (4.16) and (4.19) that $|[u_\tau]|_{bc} = 0$ at $\rho = \rho_b$. Then, since ρ^2 is a monotonically increasing function of ρ, Eq. (4.19) becomes

$$\frac{|[u_\tau]|_{bc}}{U} = \frac{\rho}{2h} - \frac{g(1)}{\rho}. \tag{4.20}$$

Using Eq. (4.4) the rate of work dissipation at the velocity discontinuity curve $0b$ is represented as

$$E_{0b} = \frac{\sigma_0}{\sqrt{3}} \iint\limits_{S_d} |[u_\tau]|_{0b} dS = \frac{2\pi HR\sigma_0}{\sqrt{3}} \int\limits_0^1 |[u_\tau]|_{0b} \sqrt{1 + \frac{1}{h^2}\left(\frac{d\rho_{0b}}{d\eta}\right)^2}\, \rho_{0b}(\eta) d\eta. \tag{4.21}$$

It follows from Eq. (4.14) that

$$\frac{d\rho_{0b}}{d\eta} = \left[\frac{\rho_{0b}(\eta)}{2} - \frac{hg(\eta)}{\rho_{0b}(\eta)}\right](1-\eta)^{-1}. \tag{4.22}$$

Then, substituting Eqs. (4.18) and (4.22) into Eq. (4.21) gives the rate of work dissipation at the velocity discontinuity curve $0b$

$$\frac{E_{0b}}{\pi U R^2 \sigma_0} = \frac{2h}{\sqrt{3}} I_1, \quad I_1 = \int_0^1 \frac{\rho_{0b}(\eta)}{(1-\eta)} \left\{ (1-\eta)^2 + \left[\frac{g(\eta)}{\rho_{0b}(\eta)} - \frac{\rho_{0b}(\eta)}{2h} \right]^2 \right\} d\eta. \quad (4.23)$$

Using Eqs. (4.4) and (4.20) the rate of work dissipation at the velocity discontinuity line bc is represented as

$$E_{bc} = \frac{\sigma_0}{\sqrt{3}} \iint_{S_d} |[u_\tau]|_{bc} dS = \frac{2\pi U R^2 \sigma_0}{\sqrt{3}} \int_{\rho_b}^1 \left[\frac{\rho^2}{2h} - g(1) \right] d\rho$$

or, after integrating,

$$\frac{E_{bc}}{\pi U R^2 \sigma_0} = \frac{1 - 3\rho_b^2 + 2\rho_b^3}{3\sqrt{3}h}. \quad (4.24)$$

Here $g(1)$ has been excluded by means of Eq. (4.16).

The rate at which forces F do work is

$$\iint_{S_v} (t_i v_i) dS = FU. \quad (4.25)$$

It has been taken into account here that two equal forces act and one half of the specimen is under consideration. Using Eq. (4.25) the inequality (1.4) for the problem under consideration can be transformed to $F_u U = E_V + E_{bc} + E_{0b}$. Substituting Eqs. (4.17), (4.23) and (4.24) into this equation it is possible to arrive at

$$f_u = \frac{F_u}{\pi R^2 \sigma_0} = \frac{2h}{\sqrt{3}} I_1 + I_2 + \frac{1 - 3\rho_b^2 + 2\rho_b^3}{3\sqrt{3}h}. \quad (4.26)$$

It is now necessary to specify the function $g(\eta)$. By analogy to Eq. (2.29) it can be chosen in the form

$$g(\eta) = \beta_0 + \beta_1 \sqrt{1 - \eta^2} \quad (4.27)$$

where β_0 and β_1 are free parameters. Using Eqs. (4.9) and (4.27) it is possible to verify by inspection that the condition (1.9) is satisfied as $\eta \to 1$. Substituting Eq. (4.27) into Eq. (4.16) shows that the limit is finite if and only if

$$\beta_0 = \frac{\rho_b^2}{2h} \quad \text{and} \quad \beta_1 = -\frac{2\rho_b^2}{\pi h}.$$

Then, replacing β_0 and β_1 in Eq. (4.27) with ρ_b and differentiating with respect to η give

$$g(\eta) = \left(1 - \frac{4}{\pi} \sqrt{1 - \eta^2} \right) \frac{\rho_b^2}{2h}, \quad g'(\eta) = \frac{2\rho_b^2 \eta}{\pi h \sqrt{1 - \eta^2}}. \quad (4.28)$$

Substituting Eq. (4.28) into Eq. (4.15) leads to

$$\rho_{0b}(\eta) = \rho_b \sqrt{\frac{1}{(1-\eta)} \left(\frac{2\eta}{\pi} \sqrt{1-\eta^2} + \frac{2}{\pi} \arcsin \eta - \eta \right)}. \qquad (4.29)$$

In general, excluding $g(\eta)$, $g'(\eta)$ and $\rho_{0b}(\eta)$ by means of Eqs. (4.28) and (4.29) the right hand side of Eq. (4.26) can be found using Eqs. (4.17) and (4.23). However, the integrals I_1 and I_2 are improper since the integrands approach infinity as $\eta \to 1$. Another difficulty is that the function $\rho_{0b}(\eta)$ reduces to the expression $0/0$ at $\eta = 1$, as follows from Eq. (4.29). In order to overcome these difficulties and facilitate numerical integration, it is convenient to introduce the new variable ϑ by

$$\eta = \sin \vartheta, \quad \sqrt{1 - \eta^2} = \cos \vartheta, \quad d\eta = \cos \vartheta d\vartheta. \qquad (4.30)$$

Then, Eqs. (4.28) and (4.29) become

$$G(\vartheta) = g[\eta(\vartheta)] = \frac{\rho_b^2}{2h} \left(1 - \frac{4}{\pi} \cos \vartheta \right), \quad G_1(\vartheta) = g'[\eta(\vartheta)] = \frac{2\rho_b^2}{\pi h} \tan \vartheta,$$

$$\Upsilon_{0b}(\vartheta) = \rho_{0b}[\eta(\vartheta)] = \rho_b \sqrt{\frac{1}{(1 - \sin \vartheta)} \left(\frac{\sin 2\vartheta}{\pi} + \frac{2\vartheta}{\pi} - \sin \vartheta \right)}. \qquad (4.31)$$

Using Eqs. (4.30) and (4.31) the integral I_1 introduced in Eq. (4.23) can be transformed to

$$I_1 = \int_0^{\pi/2} \frac{\Upsilon_{0b}(\vartheta)}{(1 - \sin \vartheta)} \left\{ (1 - \sin \vartheta)^2 + \left[\frac{G(\vartheta)}{\Upsilon_{0b}(\vartheta)} - \frac{\Upsilon_{0b}(\vartheta)}{2h} \right]^2 \right\} \cos \vartheta d\vartheta. \qquad (4.32)$$

Using Eq. (4.31) and expanding the integrand in a series in the vicinity of point $\vartheta = \pi/2$ give

$$\frac{\Upsilon_{0b}(\vartheta)}{(1 - \sin \vartheta)} \left\{ (1 - \sin \vartheta)^2 + \left[\frac{G(\vartheta)}{\Upsilon_{0b}(\vartheta)} - \frac{\Upsilon_{0b}(\vartheta)}{2h} \right]^2 \right\} \cos \vartheta$$

$$= \frac{8\rho_b^3}{9\pi^2 h^2} \left(\frac{\pi}{2} - \vartheta \right) + O\left[\left(\frac{\pi}{2} - \vartheta \right)^2 \right], \quad \vartheta \to \frac{\pi}{2}.$$

Then, I_1 can be approximated by

$$I_1 = \int_0^{\pi/2-\delta} \frac{\Upsilon_{0b}(\vartheta)}{(1 - \sin \vartheta)} \left\{ (1 - \sin \vartheta)^2 + \left[\frac{G(\vartheta)}{\Upsilon_{0b}(\vartheta)} - \frac{\Upsilon_{0b}(\vartheta)}{2h} \right]^2 \right\} \cos \vartheta d\vartheta + \frac{4\rho_b^3 \delta^2}{9\pi^2 h^2}$$

$$\qquad (4.33)$$

where $\delta \ll 1$. The integral here can be evaluated numerically with no difficulty. Using Eqs. (4.30) and (4.31) the integral I_2 introduced in Eq. (4.17) can be rewritten as

$$I_2 = 2 \int_0^{\pi/2} \int_{\Upsilon_{0b}(\vartheta)}^1 \sqrt{\frac{A(\vartheta)}{\rho^4} + \frac{B(\vartheta)}{\rho^2} + \cos^2 \vartheta}\, \rho d\rho d\vartheta,$$

$$A(\vartheta) = \frac{\rho_b^4 (1 - 4\cos\vartheta/\pi)^2 \cos^2 \vartheta}{3}, \quad B(\vartheta) = \frac{4\rho_b^4 \sin^2 \vartheta}{3\pi^2 h^2}. \tag{4.34}$$

To facilitate numerical integration, it is advantageous to integrate with respect to ρ analytically, even though the final expression is cumbersome. As a result,

$$I_2 = \int_0^{\pi/2} \left\{ \Phi(1, \vartheta) - \Phi\left[\Upsilon_{0b}^2(\vartheta),\, \vartheta\right] \right\} d\vartheta,$$

$$\Phi(\upsilon, \vartheta) = \Lambda(\upsilon, \vartheta) + \frac{B(\vartheta)}{2\cos\vartheta} \ln\left[\frac{B(\vartheta)}{\cos\vartheta} + 2\upsilon\cos\vartheta + 2\Lambda(\upsilon, \vartheta)\right]$$

$$- \sqrt{A(\vartheta)} \ln\left[\frac{2A(\vartheta) + \upsilon B(\vartheta)}{\upsilon A^{3/2}(\vartheta)} + \frac{2\Lambda(\upsilon, \vartheta)}{\upsilon A(\vartheta)}\right]. \tag{4.35}$$

where $\Lambda(\upsilon, \vartheta) = \sqrt{\upsilon^2 \cos^2 \vartheta + \upsilon B(\vartheta) + A(\vartheta)}$. It follows from this equation that $\Phi(\upsilon, \vartheta) \to \infty$ as $\vartheta \to \pi/2$. However,

$$\lim_{\vartheta \to \pi/2} [\Phi(1, \vartheta) - \Phi(\upsilon, \vartheta)] = \frac{4\rho_b^2}{\sqrt{3}\pi h}\left(1 - \sqrt{\upsilon}\right). \tag{4.36}$$

Moreover,

$$\Phi(1, \vartheta) - \Phi(\upsilon, \vartheta) = \frac{4\rho_b^2}{\sqrt{3}\pi h}\left(1 - \sqrt{\upsilon}\right) + O\left[\left(\frac{\pi}{2} - \vartheta\right)^2\right], \quad \vartheta \to \frac{\pi}{2}.$$

Therefore, using Eqs. (4.35) and (4.36) the value of I_2 can be approximated by

$$I_2 = \frac{4\rho_b^2}{\sqrt{3}\pi h}(1 - \rho_b)\delta + \int_0^{\pi/2-\delta} \left\{ \Phi(1, \vartheta) - \Phi\left[\Upsilon_{0b}^2(\vartheta),\, \vartheta\right] \right\} d\vartheta \tag{4.37}$$

where $\delta \ll 1$. Substituting Eqs. (4.33) and (4.37) into Eq. (4.26) the value of f_u can be found numerically with no difficulty. This value depends on one parameter, ρ_b. In order to find the best upper bound based on the kinematically admissible velocity field chosen, it is necessary to minimize f_u with respect to this parameter. As a result of numerical minimization, the values of f_u and ρ_b are obtained. This value of ρ_b found should be used to control the inequality $\rho_b \leq 1$. The variation of f_u with h in the range $0.05 \leq h \leq 0.5$ is depicted in Fig. 4.3. In this range of h,

Fig. 4.3 Variation of the dimensionless upper bound limit load with h

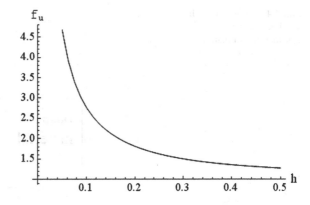

the value of ρ_b changes from 0.247 to 0.676 (approximately). It has been assumed in this numerical solution that $\delta = 10^{-4}$.

4.2 Crack at the Mid-Plane of the Weld

The geometry of the specimen and the cylindrical coordinate system are illustrated in Fig. 4.4. An axisymmetric crack is located at the mid-plane of the weld. Its radius is a_0. The problem is symmetric relative to the plane $z = 0$. Therefore, it is sufficient to find the solution in the domain $z \geq 0$. The velocity boundary conditions (4.2) and (4.3) are valid. The condition (4.1) is also valid but over the range $a_0 \leq r \leq R$. The general structure of the kinematically admissible velocity field chosen is shown in Fig. 4.5 (one half of the cross-section of the weld by a plane $\theta = $ constant is presented). The rigid zone moves along the z-axis along with the block of rigid base material. Therefore, the boundary condition (4.3) is automatically satisfied. The velocity field given in Eqs. (4.5) and (4.8) as well as the function $g(\eta)$ adopted in the previous solution (see Eq. (4.27)) can be used in the plastic zone, though the coefficients β_0 and β_1 have to be recalculated. The velocity discontinuity curve bd must contain the crack tip. Therefore, the boundary condition to Eq. (4.14) is $\rho = a$ for $\eta = 0$ where $a = a_0/R$. The solution to Eq. (4.14) satisfying this boundary condition is

$$\rho = \rho_{bk}(\eta) = \frac{1}{\sqrt{1-\eta}} \sqrt{-2h \int_0^{\eta} g(\upsilon)d\upsilon + a^2}. \qquad (4.38)$$

Equation (4.16) is valid but the admissible range for ρ_b is $a < \rho_b \leq 1$. Moreover, the limit is finite if and only if $\beta_0 = a^2/(2h) - \pi\beta_1/4$. It also follows from Eq. (4.16) that $\beta_0 = \rho_b^2/(2h)$. Then, excluding β_0 and β_1 in Eq. (4.28) leads to

Fig. 4.4 Geometry of the specimen under consideration–notation

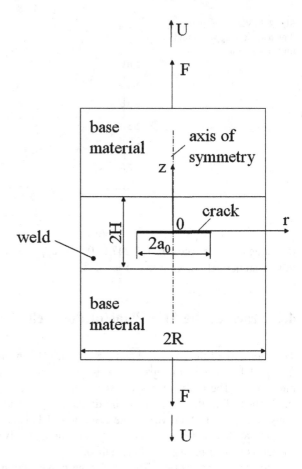

Fig. 4.5 General structure of the kinematically admissible velocity field

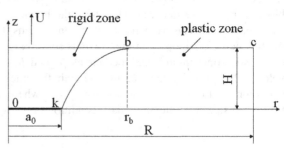

$$g(\eta) = \frac{\rho_b^2}{2h} + \frac{2}{\pi h}\left(a^2 - \rho_b^2\right)\sqrt{1 - \eta^2}, \quad g'(\eta) = -\frac{2}{\pi h}\left(a^2 - \rho_b^2\right)\frac{\eta}{\sqrt{1 - \eta^2}}. \quad (4.39)$$

Equation (4.26) for the dimensionless upper bound limit load f_u is valid but $\rho_{0b}(\eta)$ should be replaced with $\rho_{bk}(\eta)$ given in Eq. (4.38) and Eq. (4.39) should be

Fig. 4.6 Variation of a_{cr} with h

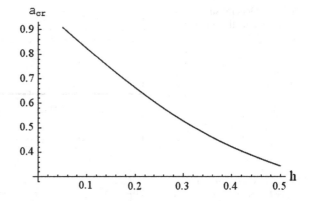

Fig. 4.7 Variation of the dimensionless upper bound limit load with a at several values of h

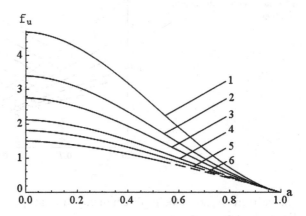

used to exclude $g(\eta)$ and $g'(\eta)$. Then, the final expression for f_u includes two geometric parameters, h and a. As before, η can be replaced with ϑ according to Eq. (4.30) and, then, f_u can be minimized with respect to r_b numerically with no difficulty. The condition $r_b \leq 1$ is not satisfied for large cracks, $a > a_{cr}$. The variation of a_{cr} with h is depicted in Fig. 4.6. The dependence of f_u on a at several values of h in the range $0 \leq a \leq a_{cr}$ is shown by solid lines in Fig. 4.7 (curve 1 corresponds to $h = 0.05$, curve 2 to $h = 0.075$, curve 3 to $h = 0.1$, curve 4 to $h = 0.15$, curve 5 to $h = 0.2$, and curve 6 to $h = 0.3$).

For specimens with $a > a_{cr}$ the general structure of the kinematically admissible velocity field is illustrated in Fig. 4.8. The length of the velocity discontinuity curve bk is now controlled by the surface $\rho = 1$ (or $r = R$) rather than $\eta = 1$ (or $\vartheta = \pi/2$). Let ϑ_d be the value of ϑ corresponding to point d (Fig. 4.8). Substituting Eq. (4.39) into Eq. (4.38), using Eq. (4.30) and putting $\rho = 1$ lead to

$$\rho_b = \sqrt{\frac{\pi a^2 - \pi(1 - \sin\vartheta_d) - a^2(2\vartheta_d + \sin 2\vartheta_d)}{\pi \sin\vartheta_d - 2\vartheta_d - \sin 2\vartheta_d}}. \tag{4.40}$$

Fig. 4.8 General structure of the kinematically admissible velocity field

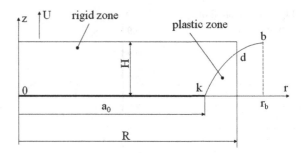

Fig. 4.9 Variation of the upper bound limit load with a for lager cracks at several h-values

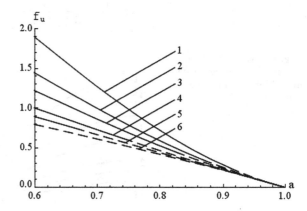

Since there is no velocity discontinuity line at $\eta = 1$ in the present solution, Eq. (4.26) transforms to

$$f_u = \frac{F_u}{\pi R^2 \sigma_0} = \frac{2h}{\sqrt{3}} I_1 + I_2. \tag{4.41}$$

The expressions for I_1 and I_2 given in Eqs. (4.17) and (4.23) are in general valid. However, η should be replaced with ϑ by means of Eq. (4.30); $\rho_{bd}(\eta)$, $g(\eta)$ and $g'(\eta)$ by functions of ϑ by means of Eqs. (4.30), (4.38) and (4.39); and ρ_b in these functions should be excluded by means of Eq. (4.40). Finally, the limits of integration with respect to ϑ in Eqs. (4.17) and (4.23) should be from 0 to ϑ_d. Then, the value of f_u found from Eq. (4.41) depends on ϑ_d. Minimizing with respect to this parameter gives the best upper bound on F based on the kinematically admissible velocity field proposed. The dependence of f_u on a at several values of h in the range $a_{cr} \le a \le 1$ is shown in Fig. 4.7 by broken lines. For larger cracks the same dependence is depicted in Fig. 4.9, including the solution for $a \le a_{cr}$ (curve 1 corresponds to $h = 0.05$, curve 2 to $h = 0.075$, curve 3 to $h = 0.1$, curve 4 to $h = 0.15$, curve 5 to $h = 0.2$, and curve 6 to $h = 0.3$).

A simpler solution for the specimen considered in this section has been proposed in Alexandrov et al. (1999).

Fig. 4.10 Geometry of the
specimen under
consideration–notation

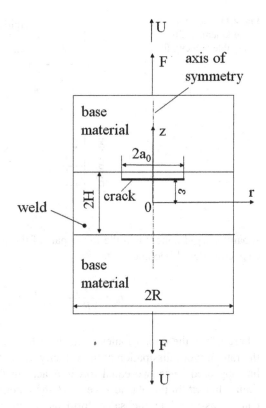

4.3 Crack at Some Distance From the Mid-Plane of the Weld

The geometry of the specimen and the cylindrical coordinate system are illustrated in Fig. 4.10. The plane $z = 0$ coincides with the plane of symmetry of the specimen with no crack. An axisymmetric crack is located within the weld at some distance ε from its mid-plane. The plane of the crack is parallel to the plane $z = 0$. It is possible to assume, with no loss of generality, that $0 \leq \varepsilon \leq H$. The blocks of rigid base material move with a velocity U along the z-axis in the opposite directions. The specimen considered in the previous section is obtained as a particular case at $\varepsilon = 0$. An important case of the interface crack is obtained at $\varepsilon = H$.

The general structure of the kinematically admissible velocity field chosen is illustrated in Fig. 4.11 (cross-section of the weld by a plane $\theta = \text{constant}$ is shown). There are two rigid zones and one plastic zone. The plastic zone $cbkb_1c_1$ is symmetric relative to the plane $z = 0$. Therefore, it is sufficient to consider its upper part $kbcd$. The rigid zones move along the z-axis in the opposite directions along with the blocks of rigid base material. The velocity discontinuity line kf separates these zones. The velocity discontinuity curve kb separates rigid zone 1 and the upper part of the plastic zone. The velocity discontinuity curve kb_1

Fig. 4.11 General structure of the kinematically admissible velocity field

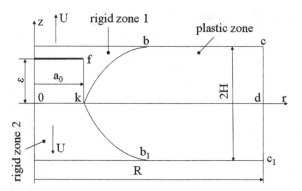

separates rigid zone 2 and the lower part of the plastic zone. Equation (1.4) in the case considered becomes

$$f_u = \frac{F_u}{\pi R^2 \sigma_0} = f_u^{(0)} + \frac{E_{kf}}{2\pi U R^2 \sigma_0} \qquad (4.42)$$

where $f_u^{(0)}$ is the dimensionless limit load for the specimen with $\varepsilon = 0$ and E_{kf} is the rate of work dissipation at the velocity discontinuity line kf. The multiplier $1/2$ has appeared since two equal forces F act but there is just one velocity discontinuity line kf. In the plastic zone $kbcd$, the kinematically admissible velocity field can be assumed in the same form as in the plastic zone shown in Fig. 4.5. Therefore, the solution presented in the previous section provides the value of $f_u^{(0)}$ (see Fig. 4.7 where f_u should be replaced with $f_u^{(0)}$). The amount of velocity jump across the velocity discontinuity line kf is equal to $|[u_\tau]|_{kf} = 2U$. Therefore, the rate of work dissipation at this line is given by

$$E_{ef} = \frac{\sigma_0}{\sqrt{3}} \iint\limits_{S_d} |[u_\tau]|_{kf} dS = \frac{4\pi U a_0 \varepsilon \sigma_0}{\sqrt{3}}. \qquad (4.43)$$

Then, it follows from Eqs. (4.42) and (4.43) that

$$f_u = f_u^{(0)} + \frac{2a\varepsilon}{\sqrt{3}R}. \qquad (4.44)$$

Thus having the value of the limit load for the specimen with a crack located at the mid-plane of the weld this simple formula can be used to find the upper bound limit load when a crack of the same radius is located at some distance from the mid-plane of the weld.

Fig. 4.12 Geometry of the
specimen under
consideration–notation

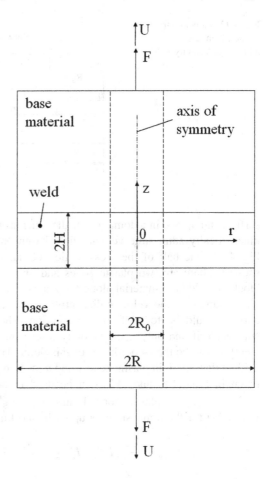

4.4 Hollow Cylinder with No Crack

The geometry of the specimen and the cylindrical coordinate system are illustrated in Fig. 4.12. The problem is symmetric relative to the plane $z = 0$. Therefore, it is sufficient to consider the domain $z \geq 0$. The velocity boundary conditions (4.1) and (4.2) are valid. The boundary condition (4.3) at $r = 0$ is replaced by the condition that the surface $r = R_0$ is traction free. In contrast to the solid cylinder, the direction of the radial velocity is not dictated by the constraints imposed. Moreover, it is very well known from the solution for upsetting of a ring between parallel plates that $u_r > 0$ in a domain near the outer surface of the ring and $u_r < 0$ in a domain near its inner surface if the friction stress is high enough (Avitzur 1980). The conditions at velocity discontinuity surfaces are mathematically equivalent to those at maximum friction surfaces in metal forming (except that the normal stress must be negative when the friction law applies). Therefore, it is reasonable to expect that in tension of the ring $u_r < 0$ in a domain near the outer

Fig. 4.13 General structure
of the kinematically
admissible velocity field

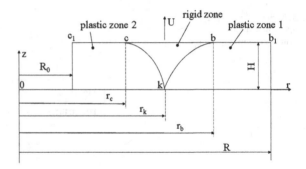

surface and $u_r > 0$ in a domain near the inner surface. The general structure of the
kinematically admissible velocity field based on this hypothesis is illustrated in
Fig. 4.13 (one half of the cross-section of the weld by a plane $\theta = $ constant is
shown). There are two plastic zones and one rigid zone moving along with the
block of rigid base material along the z-axis. The rigid zone is separated from the
plastic zones by the velocity discontinuity curves bk and ck. The shape of these
curves should be found from the solution. The velocity field is kinematically
admissible if and only if the velocity discontinuity curves bk and ck intersect at
$z = 0$. Let r_k be the r-coordinate of this point. Then, by definition, $r_k = \rho_k R$. It is
evident that $u_r < 0$ in plastic zone 1 and $u_r > 0$ in plastic zone 2. The velocity field
given in Eqs. (4.5) and (4.8) can be used in both plastic zones assuming that
$g(\eta) = g^{(1)}(\eta)$ in plastic zone 1 and $g(\eta) = g^{(2)}(\eta)$ in plastic zone 2. Then,
Eq. (4.26) for the dimensionless upper bound limit load transforms to

$$f_u = \frac{2h}{\sqrt{3}} \left(I_1^{(1)} + I_1^{(2)} \right) + I_2^{(1)} + I_2^{(2)} + I_3^{(1)} + I_3^{(2)} \qquad (4.45)$$

where $h = H/R$, $I_1^{(1)}$ is the rate of work dissipation in plastic zone 1, $I_1^{(2)}$ is the rate
of work dissipation in plastic zone 2, $I_2^{(1)}$ is the rate of work dissipation at the
velocity discontinuity curve bk, $I_2^{(2)}$ is the rate of work dissipation at the velocity
discontinuity curve ck, $I_3^{(1)}$ is the rate of work dissipation at the velocity discontinuity line bb_1, and $I_3^{(2)}$ is the rate of work dissipation at the velocity discontinuity
line cc_1. As before, the velocity discontinuity lines bb_1 and cc_1 appear because
$u_r = 0$ at $z = H$ according to the boundary condition (4.3) but the radial velocity
from Eq. (4.8) does not vanish at $z = H$.

It is convenient to replace η with ϑ using Eq. (4.30). Then, instead of the
function $g(\eta)$ and its derivative, it is necessary to use the functions $G(\vartheta)$ and
$G_1(\vartheta)$ introduced in Eq. (4.31). Let r_b and r_c be the r-coordinates of points b and c,
respectively (Fig. 4.13). Then, by definition, $r_b = \rho_b R$ and $r_c = \rho_c R$. The kinematically admissible velocity field proposed is valid if and only if

$$\rho_0 \le \rho_c, \quad 1 \ge \rho_b. \tag{4.46}$$

Consider plastic zone 1. The boundary condition to Eq. (4.14) is $\rho = \rho_k$ for $\eta = 0$. The solution to Eq. (4.14) satisfying this boundary condition and giving the equation for the velocity discontinuity curve bk is

$$\rho = \rho_{bk}(\eta) = \frac{1}{\sqrt{1-\eta}} \sqrt{-2h \int_0^\eta g^{(1)}(\upsilon) d\upsilon + \rho_k^2}. \tag{4.47}$$

Using the same arguments which have led to Eq. (4.39) the function $g^{(1)}(\eta)$ can be chosen in the form

$$g^{(1)}(\eta) = \frac{\rho_b^2}{2h} + \frac{2}{\pi h}(\rho_k^2 - \rho_b^2)\sqrt{1-\eta^2}. \tag{4.48}$$

Substituting Eq. (4.48) into Eq. (4.47) it is possible to verify by inspection that $\lim_{\eta \to 1} \rho_{bk}(\eta) = \rho_b$. Using Eqs. (4.47) and (4.48) the functions introduced in Eq. (4.31) become

$$G^{(1)}(\vartheta) = g^{(1)}[\eta(\vartheta)] = \frac{\rho_b^2}{2h} + \frac{2}{\pi h}(\rho_k^2 - \rho_b^2)\cos\vartheta,$$

$$\Upsilon_{bk}(\vartheta) = \rho_{bk}[\eta(\vartheta)] = \frac{1}{\sqrt{1-\sin\vartheta}}\sqrt{\rho_k^2 - \rho_b^2 \sin\vartheta + \frac{(\rho_b^2 - \rho_k^2)}{\pi}(2\vartheta + \sin 2\vartheta)}. \tag{4.49}$$

Plastic zone 2 can be treated in a similar manner. As a result,

$$G^{(2)}(\vartheta) = g^{(2)}[\eta(\vartheta)] = \frac{\rho_c^2}{2h} + \frac{2}{\pi h}(\rho_k^2 - \rho_c^2)\cos\vartheta,$$

$$\Upsilon_{ck}(\vartheta) = \rho_{ck}[\eta(\vartheta)] = \frac{1}{\sqrt{1-\sin\vartheta}}\sqrt{\rho_k^2 - \rho_c^2 \sin\vartheta + \frac{(\rho_c^2 - \rho_k^2)}{\pi}(2\vartheta + \sin 2\vartheta)}. \tag{4.50}$$

The integrals $I_1^{(1)}$ and $I_1^{(2)}$ involved in Eq. (4.45) become

$$I_1^{(1)} = \int_0^{\pi/2} \frac{\Upsilon_{bk}(\vartheta)}{(1-\sin\vartheta)} \left\{ (1-\sin\vartheta)^2 + \left[\frac{G^{(1)}(\vartheta)}{\Upsilon_{bk}(\vartheta)} - \frac{\Upsilon_{bk}(\vartheta)}{2h}\right]^2 \right\} \cos\vartheta d\vartheta \tag{4.51}$$

$$I_1^{(2)} = \int_0^{\pi/2} \frac{\Upsilon_{ck}(\vartheta)}{(1-\sin\vartheta)} \left\{ (1-\sin\vartheta)^2 + \left[\frac{G^{(2)}(\vartheta)}{\Upsilon_{ck}(\vartheta)} - \frac{\Upsilon_{ck}(\vartheta)}{2h}\right]^2 \right\} \cos\vartheta d\vartheta$$

where $G^{(1)}(\vartheta)$, $G^{(2)}(\vartheta)$, $\Upsilon_{bk}(\vartheta)$ and $\Upsilon_{ck}(\vartheta)$ should be excluded by means of Eqs. (4.49) and (4.50).

Equation (4.34) is in general valid to determine $I_2^{(1)}$ and $I_2^{(2)}$ except for the limits of integration with respect to ρ. In particular, Eq. (4.35) should be replaced with

$$I_2^{(1)} = \int_0^{\pi/2} \left\{ \Phi(1,\ \vartheta) - \Phi\left[\Upsilon_{bk}^2(\vartheta),\ \vartheta\right] \right\} d\vartheta,$$

$$I_2^{(2)} = \int_0^{\pi/2} \left\{ \Phi\left[\Upsilon_{ck}^2(\vartheta),\ \vartheta\right] - \Phi(\rho_0^2,\ \vartheta) \right\} d\vartheta.$$

(4.52)

The functions $A(\vartheta)$ and $B(\vartheta)$ involved in the definition for the function $\Phi(v, \vartheta)$ are now given by

$$A_1(\vartheta) = \frac{1}{3}\left[\rho_b^2 + \frac{4}{\pi}\left(\rho_k^2 - \rho_b^2\right)\cos\vartheta\right]^2 \cos^2\vartheta, \quad B_1(\vartheta) = \frac{4\left(\rho_k^2 - \rho_b^2\right)^2 \sin^2\vartheta}{3\pi^2 h^2},$$

$$A_2(\vartheta) = \frac{1}{3}\left[\rho_c^2 + \frac{4}{\pi}\left(\rho_k^2 - \rho_c^2\right)\cos\vartheta\right]^2 \cos^2\vartheta, \quad B_2(\vartheta) = \frac{4\left(\rho_k^2 - \rho_c^2\right)^2 \sin^2\vartheta}{3\pi^2 h^2}$$

(4.53)

where $A(\vartheta) = A_1(\vartheta)$ and $B(\vartheta) = B_1(\vartheta)$ in plastic zone 1, $A(\vartheta) = A_2(\vartheta)$ and $B(\vartheta) = B_2(\vartheta)$ in plastic zone 2.

The amounts of velocity jump across the velocity discontinuity lines bb_1 and cc_1 are equal to $|u_r|$ at $\eta = 1$. Then, taking into account the general expression for the radial velocity given in Eq. (4.8) and the transformation Eq. (4.30) the amounts of velocity jump across the velocity discontinuity lines bb_1 and cc_1 are determined using Eqs. (4.49) and (4.50) as

$$|[u_\tau]|_{bb_1} = \frac{U}{2h}\frac{\left(\rho^2 - \rho_b^2\right)}{\rho}, \quad |[u_\tau]|_{cc_1} = \frac{U}{2h}\frac{\left(\rho_c^2 - \rho^2\right)}{\rho}. \tag{4.54}$$

Then, the integrals $I_3^{(1)}$ and $I_3^{(2)}$ involved in Eq. (4.45) are obtained in the form

$$I_3^{(1)} = \frac{1 - 3\rho_b^2 + 2\rho_b^3}{3\sqrt{3}h}, \quad I_3^{(2)} = \frac{\rho_0^3 - 3\rho_0\rho_c^2 + 2\rho_c^3}{3\sqrt{3}h}. \tag{4.55}$$

Substituting Eqs. (4.51), (4.52) and (4.55) into Eq. (4.45) yields the value of f_u as a function of three parameters, ρ_b, ρ_c and ρ_k. Minimizing f_u with respect to these parameters numerically gives the best upper bound limit load based on the velocity field proposed. This solution provides the values of ρ_b and ρ_c at which f_u attains a minimum. It is necessary to check that these values satisfy the inequalities (4.46). No numerical solution is available at the time of writing.

One (or both) of the inequalities (4.46) is not satisfied over a certain range of geometric parameters. In this case, the solution can be modified using an approach similar to that used to modify the solution (4.26) to arrive at the solution (4.41).

References

S.E. Alexandrov, R.V. Goldstein, N.N. Tchikanova, Upper bound limit load solutions for a round welded bar with an internal axisymmetric crack. Fat. Fract. Eng. Mater. Struct. **22**, 775–780 (1999)

B. Avitzur, *Metal Forming: The Application of Limit Analysis* (Dekker, New York, 1980)

Chapter 5
Plane Strain Solutions for Highly Undermatched Specimens in Bending

This chapter concerns with highly undermatched welded plates with no crack subject to pure bending. Two different kinematically admissible velocity fields are adopted. One of the fields results from an exact analytic solution for deformation of a plastic wedge between two rotating plates. The other kinematically admissible velocity field is obtained using simple assumptions and the known asymptotic singular behaviour of the real velocity field near velocity discontinuity surfaces given in Eq. (1.9). Since plane strain deformation is assumed, integration in the thickness direction is replaced with the multiplier $2B$ where $2B$ is the thickness of the plate. For the same reason, the term "velocity discontinuity surface" is replaced with the term "velocity discontinuity curve (or line)". The latter refers to curves (lines) in the plane of flow. Base material is supposed to be rigid. The solutions given below have been proposed by Alexandrov and Kocak (2007) and Alexandrov (2008). Reviews of other solutions, including plates with cracks and three-point bending, have been presented in Kim and Schwalbe (2001a, b, c).

5.1 Deformation of a Plastic Wedge Between Rotating Plates

Consider plane strain deformation of rigid perfectly/plastic material between two plates inclined at an angle 2α and which intersect in a line. The line of intersection is a hinge and the plates rotate around this hinge with an angular velocity ω in the opposite directions, as illustrated in Fig. 5.1. It is convenient to introduce a cylindrical coordinate system (r, θ, z) with its z-axis being perpendicular to the plane of flow. Also, the z-axis coincides the axis of rotation of the plates. The flow is symmetric relative to the axis $\theta = 0$. It is therefore sufficient to obtain the solution in the region $0 \leq \theta \leq \alpha$. It is assumed that there is no material flux at $r = 0$. The material sticks at the plates independently of the sense of the normal stress.

The system of equations consists of the equilibrium equations

S. Alexandrov, *Upper Bound Limit Load Solutions for Welded Joints with Cracks*, SpringerBriefs in Computational Mechanics, DOI: 10.1007/978-3-642-29234-7_5,

Fig. 5.1 Geometry of the
process

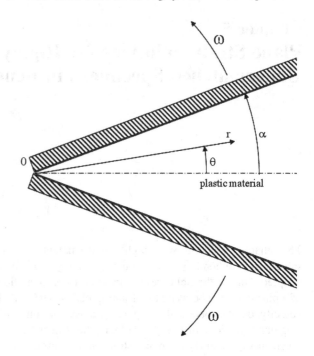

The non-zero components of the equilibrium equations

$$\frac{\partial \sigma_{rr}}{\partial r} + \frac{1}{r}\frac{\partial \sigma_{r\theta}}{\partial \theta} + \frac{\sigma_{rr} - \sigma_{\theta\theta}}{r} = 0, \qquad \frac{1}{r}\frac{\partial \sigma_{\theta\theta}}{\partial \theta} + \frac{\partial \sigma_{r\theta}}{\partial r} + \frac{2\sigma_{r\theta}}{r} = 0, \qquad (5.1)$$

the yield criterion

$$(\sigma_{rr} - \sigma_{\theta\theta})^2 + 4\sigma_{r\theta}^2 = \frac{4}{3}\sigma_0^2 \qquad (5.2)$$

and its associated flow rule

$$\xi_{rr} = \lambda(\sigma_{rr} - \sigma), \quad \xi_{\theta\theta} = \lambda(\sigma_{\theta\theta} - \sigma), \quad \xi_{r\theta} = \lambda\sigma_{r\theta}. \qquad (5.3)$$

Here σ_{rr}, $\sigma_{\theta\theta}$ and $\sigma_{r\theta}$ are the components of the stress tensor in the cylindrical coordinates, ξ_{rr}, $\xi_{\theta\theta}$ and $\xi_{r\theta}$ are the components of the strain rate tensor in the cylindrical coordinates, λ is a non-negative multiplier, σ is the hydrostatic stress defined by

$$2\sigma = \sigma_{rr} + \sigma_{\theta\theta}. \qquad (5.4)$$

Under plane strain conditions the axial strain rate is one of the principal strain rates and it vanishes everywhere, $\xi_{zz} = 0$. Therefore, the incompressibility Eq. (1.6) reduces to $\xi_{rr} + \xi_{\theta\theta} = 0$. This equation immediately follows from Eqs. (5.3) and (5.4) as well. The non-zero components of the strain rate tensor are expressed in terms of the radial and circumferential velocities, u_r and u_θ, as

$$\xi_{rr} = \frac{\partial u_r}{\partial r}, \quad \xi_{\theta\theta} = \frac{1}{r}\frac{\partial u_\theta}{\partial \theta} + \frac{u_r}{r}, \quad \xi_{r\theta} = \frac{1}{2}\left(\frac{1}{r}\frac{\partial u_r}{\partial \theta} + \frac{\partial u_\theta}{\partial r} - \frac{u_\theta}{r}\right). \tag{5.5}$$

The yield criterion (5.2) is satisfied by the standard substitution

$$\sigma_{rr} = \sigma + \frac{\sigma_0}{\sqrt{3}}\cos 2\varphi, \quad \sigma_{\theta\theta} = \sigma - \frac{\sigma_0}{\sqrt{3}}\cos 2\varphi, \quad \sigma_{r\theta} = \frac{\sigma_0}{\sqrt{3}}\sin 2\varphi. \tag{5.6}$$

The direction of rotation of the plates dictates that $\xi_{\theta\theta} \geq 0$ and, therefore, the incompressibility equation gives $\xi_{rr} \leq 0$. Hence, $\xi_{rr} - \xi_{\theta\theta} \leq 0$ and it follows from Eq. (5.3) that $\sigma_{rr} - \sigma_{\theta\theta} \leq 0$. The direction of rotation of the plates and the condition of no material flux at $r = 0$ dictate that $u_r < 0$. Therefore, $\sigma_{r\theta} \geq 0$. Using these inequalities for stress components it is possible to conclude from Eq. (5.6) that

$$\pi/4 \leq \varphi \leq \pi/2. \tag{5.7}$$

It is assumed that the line $\theta = \alpha$ is a velocity discontinuity line. Then, the stress boundary conditions are $\sigma_{r\theta} = 0$ at $\theta = 0$ and $\sigma_{r\theta} = \sigma_0/\sqrt{3}$ at $\theta = \alpha$. Taking into account Eqs. (5.6) and (5.7) these boundary conditions can be transformed to

$$\varphi = \pi/2 \tag{5.8}$$

at $\theta = 0$ and

$$\varphi = \pi/4 \tag{5.9}$$

at $\theta = \alpha$.

The main assumption is that φ is independent of r. Then, substituting Eq. (5.6) into Eq. (5.1) yields

$$r\frac{\partial \sigma}{\partial r} + \frac{2\sigma_0}{\sqrt{3}}\cos 2\varphi\left(\frac{d\varphi}{d\theta} + 1\right) = 0, \quad \frac{\partial \sigma}{\partial \theta} + \frac{2\sigma_0}{\sqrt{3}}\sin 2\varphi\left(\frac{d\varphi}{d\theta} + 1\right) = 0. \tag{5.10}$$

These equations are compatible if and only if

$$\frac{\sigma}{\sigma_0} = \frac{A}{\sqrt{3}}\ln r + p_0(\theta) \tag{5.11}$$

where A is constant and p_0 is an arbitrary function of θ. Substituting Eq. (5.11) into the first of Eq. (5.10) gives

$$\frac{d\varphi}{d\theta} = -\frac{(A + 2\cos 2\varphi)}{2\cos 2\varphi}. \tag{5.12}$$

The solution to this equation satisfying the boundary conditions (5.8) and (5.9) determines φ as a function of θ and the value of A. Substituting this solution and Eq. (5.11) into the second of Eq. (5.10) it is possible to find the function $p_0(\theta)$. However, this function is not essential for the purpose of the present investigation

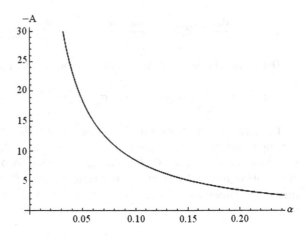

Fig. 5.2 Variation of A with α

and, therefore, will not be determined. In general, Eq. (5.12) can be integrated in elementary functions. However, the final expression is cumbersome [see, for example, the solution for planar flow of plastic material through a converging channel presented in Hill (1950)]. Therefore, it is more convenient to use the solution to Eq. (5.12) in the following form

$$\theta = -2 \int_{\pi/2}^{\varphi} \frac{\cos 2\chi}{(A + 2\cos 2\chi)} d\chi \tag{5.13}$$

where χ is a dummy variable. The solution in this form satisfies the boundary condition (5.8). Substituting Eq. (5.9) into Eq. (5.13) yields

$$\alpha = 2 \int_{\pi/4}^{\pi/2} \frac{\cos 2\chi}{(A + 2\cos 2\chi)} d\chi. \tag{5.14}$$

The solution to this equation determines A as a function of α. This function is illustrated in Fig. 5.2. Thus, the stress field found satisfies the equilibrium equations, the yield criterion and the stress boundary conditions at $\theta = 0$ and $\theta = \alpha$.

Substituting the stress field found into the associated flow rule (5.3) and using Eq. (5.5) result in the system of equations for the velocity components. The circumferential velocity can be assumed in the form

$$u_\theta = \omega r u_0(\theta) \tag{5.15}$$

where u_0 is an arbitrary function of θ or φ since φ is the function of θ determined by Eq. (5.13). The function u_0 must satisfy the following boundary conditions

$$u_0 = 0 \tag{5.16}$$

at $\theta = 0$ (or $\varphi = \pi/2$) and

$$u_0 = 1 \qquad (5.17)$$

at $\theta = \alpha$ (or $\varphi = \pi/4$). The condition (5.16) is a symmetry condition and the condition (5.17) follows from the continuity of the normal velocity across the velocity discontinuity line $\theta = \alpha$. Using Eqs. (5.5) and (5.15) the incompressibility equation $\xi_{rr} + \xi_{\theta\theta} = 0$ can be written in the form

$$\frac{\partial u_r}{\partial r} + \frac{u_r}{r} + \omega \frac{du_0}{d\theta} = 0.$$

The solution to this equation giving a finite magnitude of the radial velocity at the origin $r = 0$ is

$$u_r = -\frac{\omega r}{2} \frac{du_0}{d\theta}. \qquad (5.18)$$

Excluding λ in Eq. (5.3) yields

$$\frac{\sigma_{rr} - \sigma_{\theta\theta}}{\sigma_{r\theta}} = \frac{\xi_{rr} - \xi_{\theta\theta}}{\xi_{r\theta}}.$$

Substituting Eqs. (5.5), (5.6), (5.15), and (5.18) into this equation gives

$$\frac{1}{2} \frac{d^2 u_0}{d\theta^2} = \tan 2\varphi \frac{du_0}{d\theta}. \qquad (5.19)$$

Using Eq. (5.12) differentiation with respect to θ in this equation can be replaced with differentiation with respect to φ. Then, the solution to Eq. (5.19) satisfying the boundary conditions (5.16) and (5.17) is obtained in the following form

$$u_0 = \sin 2\varphi. \qquad (5.20)$$

The velocity components can be found from Eqs. (5.15) and (5.18), with the use of Eqs. (5.12) and (5.20), as

$$u_\theta = \omega r \sin 2\varphi, \quad u_r = \frac{\omega r}{2}(A + 2\cos 2\varphi). \qquad (5.21)$$

The components of the strain rate tensor are determined from Eqs. (5.5) and (5.21) as

$$\xi_{rr} = \frac{\omega}{2}(A + 2\cos 2\varphi), \quad \xi_{\theta\theta} = -\frac{\omega}{2}(A + 2\cos 2\varphi),$$
$$\xi_{r\theta} = \frac{\omega}{2}\tan 2\varphi(A + 2\cos 2\varphi) \qquad (5.22)$$

The solution to Eq. (5.14) illustrated in Fig. 5.2 shows that $A < 0$ and $A + 2\cos 2\varphi < 0$ at any φ of the interval (5.7). Then, the equivalent strain rate is determined from Eq. (5.22) as

$$\xi_{eq} = \sqrt{\frac{2}{3}\left(\xi_{rr}^2 + \xi_{\theta\theta}^2 + 2\xi_{r\theta}^2\right)} = \frac{\omega(A + 2\cos 2\varphi)}{\sqrt{3}\cos 2\varphi}. \tag{5.23}$$

It is evident from Eq. (5.23) that $\xi_{eq} \to \infty$ as $\varphi \to \pi/4$ (or $\theta \to \alpha$). Expanding the right hand side of this equation in a series in the vicinity of the point $\varphi = \pi/4$ leads to

$$\xi_{eq} = -\frac{\omega A}{2\sqrt{3}(\varphi - \pi/4)} + o\left[(\varphi - \pi/4)^{-1}\right], \quad \varphi \to \frac{\pi}{4}. \tag{5.24}$$

On the other hand, Eq. (5.12) can be represented as

$$\frac{d\varphi}{d\theta} = \frac{A}{4(\varphi - \pi/4)} + o\left[(\varphi - \pi/4)^{-1}\right], \quad \varphi \to \frac{\pi}{4}.$$

Integrating this equation and using the boundary condition (5.9) yield

$$\varphi - \frac{\pi}{4} = \sqrt{-\frac{A}{2}(\alpha - \theta)} \tag{5.25}$$

to leading order. Substituting Eq. (5.25) into Eq. (5.24) demonstrates that the distribution of the equivalent strain rate in the vicinity of the velocity discontinuity line $\theta = \alpha$ follows the singular asymptotic behaviour presented in Eq. (1.9).

5.2 Upper Bound Solution I

The geometry of the specimen and Cartesian coordinates (x, y) are shown in Fig. 5.3. The width of the specimen is denoted by $2W$ and the thickness of the weld by $2H$. The specimen is loaded by two bending moments G whose magnitude at plastic collapse should be found. The blocks of rigid base material rotate with an angular velocity ω around the origin of the coordinate system in the opposite directions, as illustrated in Fig. 5.3. The x-axis is perpendicular to the weld. The origin of the coordinate system is located at the intersection of the two axes of symmetry of the specimen. The rate of work dissipation in plastic zones above and below the x-axis is the same for pressure-independent materials. Therefore, it is sufficient to consider the domain $x \geq 0$ and $y \geq 0$. Let u_x be the velocity component in the x-direction and u_y in the y-direction. The velocity boundary conditions are

$$u_x = \omega y \tag{5.26}$$

at $x = H$,

$$u_x = 0 \tag{5.27}$$

at $x = 0$ and

Fig. 5.3 Geometry of the specimen–notation

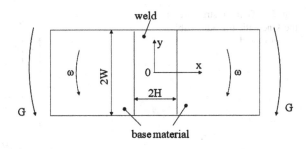

$$u_y = 0 \tag{5.28}$$

at $x = H$.

The general structure of the kinematically admissible velocity field proposed is illustrated in Fig. 5.4 (one quarter of the weld is shown). It consists of the plastic zone $0bd$ and the rigid zone $0bc$. The rigid zone rotates around the origin of the coordinate system along with the block of rigid base material. Therefore, the velocity boundary conditions (5.26) and (5.28) are automatically satisfied. It is convenient to introduce polar coordinates (r, θ) by the following transformation equations

$$x = r\sin\theta, \quad y = r\cos\theta. \tag{5.29}$$

The velocity components in the polar coordinates, u_r and u_θ, are related to u_x and u_y by

$$u_r = u_x\sin\theta + u_y\cos\theta, \quad u_\theta = -u_x\cos\theta + u_y\sin\theta. \tag{5.30}$$

The equation for the rigid plastic boundary $0b$, which is also a velocity discontinuity line, is $\theta = \alpha$ where $\tan\alpha = h$. Here and in what follows the dimensionless parameter h denotes the ratio H/W. Using Eqs. (5.29) and (5.30) the velocity boundary condition (5.27) can be transformed to

$$u_\theta = 0 \tag{5.31}$$

at $\theta = 0$. The normal velocity must be continuous across the velocity discontinuity line $0b$. Therefore,

$$u_\theta = \omega r \tag{5.32}$$

at $\theta = \alpha$. Taking into account Eq. (5.15) it is possible to conclude that the boundary conditions (5.16) and (5.17) are equivalent to the boundary conditions (5.31) and (5.32). Therefore, the velocity field (5.21) is kinematically admissible for the problem under consideration.

Using Eq. (5.23) the rate of work dissipation in the plastic zone is determined as

$$E_V = \sigma_0 \iiint_V \zeta_{eq} dV = \frac{2B\omega\sigma_0}{\sqrt{3}} \int_0^\alpha \int_0^{R_{db}(\theta)} \frac{(A + 2\cos 2\varphi)}{\cos 2\varphi} r dr d\theta. \tag{5.33}$$

Fig. 5.4 General structure of
the kinematically admissible
velocity field

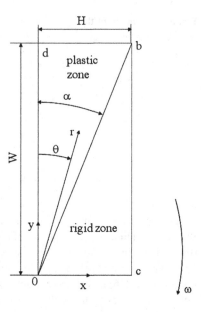

Here $r = R_{db}(\theta)$ is the equation for the line db (Fig. 5.4) in the polar coordinates. Since this line is determined by the equation $y = W$ in the Cartesian coordinates, it follows from Eq. (5.29) that

$$R_{db}(\theta) = \frac{W}{\cos\theta}. \tag{5.34}$$

Since φ is independent of r, integration with respect to this variable in Eq. (5.33) can be immediately carried out to give, with the use of Eq. (5.34),

$$E_V = \frac{BW^2\omega\sigma_0}{\sqrt{3}} \int\limits_0^\alpha \frac{(A + 2\cos 2\varphi)}{\cos^2\theta\cos 2\varphi} d\theta. \tag{5.35}$$

Using Eq. (5.12) integration with respect to θ in Eq. (5.35) can be replaced with integration with respect to φ. Then,

$$E_V = \frac{2BW^2\omega\sigma_0}{\sqrt{3}} \int\limits_{\pi/4}^{\pi/2} \frac{d\varphi}{\cos^2\theta}. \tag{5.36}$$

Equation (5.13) should be used to replace $\cos\theta$ with a function of φ here. Then, the integral in Eq. (5.36) can be evaluated numerically with no difficulty. It is evident from Eq. (5.13) that the value of E_V depends on A.

The amount of velocity jump across the velocity discontinuity line $0b$ is simply equal to $|u_r|$ at $\theta = \alpha$ [or $\varphi = \pi/4$ as follows from Eq. (5.9)]. Let R_b be the r-coordinate of point b (Fig. 5.4). Then, the rate of work dissipation at the velocity discontinuity line $0b$ is given by

$$E_{0b} = \frac{\sigma_0}{\sqrt{3}} \iint_{S_d} |[u_\tau]|_{0b} dS = \frac{2B\sigma_0}{\sqrt{3}} \int_0^{R_b} |u_r|_{\varphi=\pi/4} dr. \qquad (5.37)$$

Since $A < 0$ (Fig. 5.2), it follows from Eq. (5.21) that

$$|u_r|_{\varphi=\pi/4} = -\frac{\omega A r}{2}. \qquad (5.38)$$

It is evident from Fig. 5.4 that $R_b = W/\cos \alpha$. Therefore, substituting Eq. (5.38) into Eq. (5.37) and integrating lead to

$$E_{0b} = -\frac{\omega B W^2 A \sigma_0}{2\sqrt{3} \cos^2 \alpha}. \qquad (5.39)$$

The rate at which external forces do work is

$$\iint_{S_v} (t_i v_i) dS = \frac{1}{2} G\omega. \qquad (5.40)$$

It has been taken into account here that two equal bending moments act but one quarter of the specimen is under consideration.

Using Eq. (5.40) the inequality (1.4) for the problem under consideration can be represented as

$$g_u = \frac{G_u}{2BW^2\sigma_0} = \frac{(E_V + E_{0b})}{\omega BW^2\sigma_0}. \qquad (5.41)$$

Here G_u is the upper bound limit moment and g_u is its dimensionless representation. Substituting Eqs. (5.36) and (5.39) into (5.41) yields

$$g_u = \frac{2}{\sqrt{3}} \int_{\pi/4}^{\pi/2} \frac{d\varphi}{\cos^2 \theta} - \frac{A}{2\sqrt{3} \cos^2 \alpha}. \qquad (5.42)$$

Since the dependence of A on α has been already found (Fig. 5.2) and $\tan \alpha = h$, Eq. (5.42) provides the variation of the dimensionless upper bound bending moment with h. Eq. (5.42) contains no free parameters. Therefore, minimization is not required. The dependence of g_u on h is depicted in Fig. 5.5.

Fig. 5.5 Variation of the
dimensionless bending
moment with h

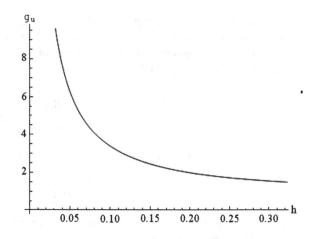

5.3 Upper Bound Solution II

The previous solution is essentially based on the exact analytic solution for
deformation of a plastic wedge between rotating plates. A more universal approach
is adopted below. It can be viewed as an extension of the approach used to arrive at
Eq. (2.28). The solution will be given in the Cartesian coordinates shown in
Fig. 5.3. The general structure of the kinematically admissible velocity field
proposed is illustrated in Fig. 5.6 (one quarter of the weld is shown). The rigid
zone rotates around the origin of the coordinate system along with the block of
rigid base material. It is separated from the plastic zone by the velocity discontinuity curve $0b$ whose shape should be found from the solution. The velocity
vector in the rigid zone is given by

$$\mathbf{U} = \omega y \mathbf{i} - \omega x \mathbf{j} \qquad (5.43)$$

where $\mathbf{i}$ and $\mathbf{j}$ are the base vectors of the Cartesian coordinates. Assume the
distribution of the velocity component u_x in the plastic zone in the form

$$u_x = \frac{\omega y x}{H}. \qquad (5.44)$$

It is evident that the boundary conditions (5.26) and (5.27) are automatically
satisfied. Also satisfied is the boundary condition (5.28) but only over the portion
of the line $y = H$ where the rigid zone sticks at the block of base material. The
equation of incompressibility (1.6) in the case under consideration reduces to
$\partial u_x / \partial x + \partial u_y / \partial y = 0$. Substituting Eq. (5.44) into this equation and integrating
give

$$u_y = -\frac{\omega y^2}{2H} + \omega H f(x) \qquad (5.45)$$

Fig. 5.6 General structure of the kinematically admissible velocity field

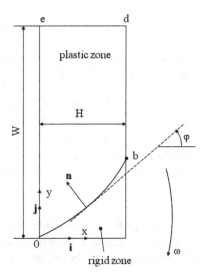

where $f(x)$ is an arbitrary function of x. Symmetry dictates that $f(x)$ is an even function of x. Then, by analogy to Eq. (2.29), one of the simplest representations of this function satisfying the singular asymptotic behaviour given in Eq. (1.9) is

$$f(x) = \beta_0 + \beta_1 \sqrt{1 - \left(\frac{x}{H}\right)^2}. \tag{5.46}$$

It is convenient to introduce the following dimensionless quantities

$$\varsigma = \frac{y}{W}, \quad \sin \vartheta = \frac{x}{H}, \quad h = \frac{H}{W}. \tag{5.47}$$

Then, using Eqs. (5.44), (5.45) and (5.46) the velocity vector in the plastic zone is represented as

$$\mathbf{u} = u_x \mathbf{i} + u_y \mathbf{j} = \omega W \left\{ \varsigma \sin \vartheta \mathbf{i} + \left[h(\beta_0 + \beta_1 \cos \vartheta) - \frac{\varsigma^2}{2h} \right] \mathbf{j} \right\}. \tag{5.48}$$

Equation (5.43) becomes

$$\mathbf{U} = \omega W (\varsigma \mathbf{i} - h \sin \vartheta \mathbf{j}). \tag{5.49}$$

Let φ be the orientation of the tangent to the velocity discontinuity curve $0b$ relative to the x-axis, measured anti-clockwise from the axis (Fig. 5.6). Then, the unit normal vector to this curve can be represented as

$$\mathbf{n} = -\sin \varphi \mathbf{i} + \cos \varphi \mathbf{j}. \tag{5.50}$$

By definition, $\tan \varphi = dy/dx$. Then, using Eq. (5.47)

$$\tan \varphi = \frac{1}{h \cos \vartheta} \frac{d\varsigma}{d\vartheta}. \tag{5.51}$$

Assume that $\mathbf{U} \equiv \mathbf{u_1}$ and $\mathbf{u} \equiv \mathbf{u_2}$ in Eq. (1.7). Then, it follows from Eqs. (5.48), (5.49), (5.50), and (5.51) that

$$\frac{2\varsigma(1 - \sin \vartheta)}{\cos \vartheta} \frac{d\varsigma}{d\vartheta} = \varsigma^2 - 2(\beta_0 + \beta_1 \cos \vartheta + \sin \vartheta)h^2. \tag{5.52}$$

The velocity field proposed is kinematically admissible if and only if the velocity discontinuity curve $0b$ contains the origin of the coordinate system. Therefore, the boundary condition to Eq. (5.52) is

$$\varsigma = 0 \tag{5.53}$$

at $\vartheta = 0$. Equation (5.52) reduces to a linear differential equation of first order for ς^2. Therefore, its general solution can be found with no difficulty. In particular, the solution satisfying the boundary condition (5.53) is

$$\varsigma = \varsigma_{0b}(\vartheta) = h\sqrt{-\frac{\sin^2 \vartheta + \beta_1(\vartheta + \sin \vartheta \cos \vartheta) + 2\beta_0 \sin \vartheta}{1 - \sin \vartheta}}. \tag{5.54}$$

In general, it follows from this solution that $\varsigma_{0b}(\vartheta) \to \infty$ as $\vartheta \to \pi/2$ (or $x \to H$). In order to get a finite value of ς at $\vartheta = \pi/2$, it is necessary to put

$$\beta_0 = -\frac{1}{2}\left(1 + \frac{\pi\beta_1}{2}\right). \tag{5.55}$$

In this case Eq. (5.54) becomes

$$\varsigma = \varsigma_{0b}(\vartheta) = \frac{h}{\sqrt{2}}\sqrt{\frac{2\sin \vartheta(1 - \sin \vartheta) - \beta_1(2\vartheta + \sin 2\vartheta - \pi \sin \vartheta)}{1 - \sin \vartheta}}. \tag{5.56}$$

The right hand side of this equation reduces to the expression $0/0$ at $\vartheta = \pi/2$. Applying l'Hospital's rule the value of ς at point b (Fig. 5.6) is determined as

$$\varsigma_b = \lim_{\vartheta \to \pi/2} \varsigma_{0b}(\vartheta) = h\sqrt{1 - \pi\beta_1/2}. \tag{5.57}$$

The strain rate components and the equivalent strain rate are found from Eqs. (1.3), (5.47), (5.48) and (5.55) as

$$\begin{aligned}
\zeta_{xx} &= \frac{\partial u_x}{\partial x} = \frac{\omega\varsigma}{h}, \quad \zeta_{yy} = \frac{\partial u_y}{\partial y} = -\frac{\omega\varsigma}{h}, \\
\zeta_{xy} &= \frac{1}{2}\left(\frac{\partial u_x}{\partial y} + \frac{\partial u_y}{\partial x}\right) = \frac{\omega \sin \vartheta}{2}\left(1 - \frac{\beta_1}{\cos \vartheta}\right), \\
\zeta_{eq} &= \frac{\omega}{\sqrt{3}h \cos \vartheta}\sqrt{4\varsigma^2 \cos^2 \vartheta + h^2 \sin^2 \vartheta(\cos \vartheta - \beta_1)^2}.
\end{aligned} \tag{5.58}$$

The rate of work dissipation in the plastic zone is determined from Eqs. (5.47) and (5.58) in the form

$$
E_V = \sigma_0 \iiint_V \zeta_{eq} dV
$$

$$
= \frac{2BW^2\omega\sigma_0}{\sqrt{3}} \int_0^{\pi/2} \int_{\varsigma_{0b}(\vartheta)}^1 \sqrt{4\varsigma^2 \cos^2\vartheta + h^2 \sin^2\vartheta(\cos\vartheta - \beta_1)^2} \, d\varsigma d\vartheta.
$$

To facilitate numerical integration, it is advantageous to integrate with respect to ς analytically, even though the final expression is cumbersome. As a result,

$$
\frac{E_V}{BW^2\omega\sigma_0} = \frac{2}{\sqrt{3}} \int_0^{\pi/2} \cos\vartheta\{\Phi(1, \vartheta) - \Phi[\varsigma_{0b}(\vartheta), \vartheta]\}d\vartheta,
$$

$$
\Phi(v, \vartheta) = v\sqrt{v^2 + \Lambda^2(\vartheta)} + \Lambda^2(\vartheta) \ln\left[v + \sqrt{v^2 + \Lambda^2(\vartheta)}\right],
$$

$$
\Lambda(\vartheta) = \frac{h}{2}\tan\vartheta(\cos\vartheta - \beta_1).
$$

(5.59)

There are two velocity discontinuity curves, $0b$ and bd (Fig. 5.6). In particular, the velocity discontinuity line bd appears because $u_x = 0$ at $x = H$ according to the boundary condition (5.28) but $u_x \neq 0$ at $x = H$ according to Eq. (5.45). Substituting Eqs. (5.48) and (5.49) into Eq. (1.8) gives the amount of velocity jump across the velocity discontinuity curve $0b$ in the form

$$
|[u_\tau]|_{0b} = \omega W \sqrt{(1 - \sin\vartheta)^2 \varsigma_{0b}^2(\vartheta) + \left[(\beta_0 + \beta_1 \cos\vartheta + \sin\vartheta)h - \frac{\varsigma_{0b}^2(\vartheta)}{2h}\right]^2}.
$$

(5.60)

Using Eq. (5.47) the rate of work dissipation at this line can be represented in the following form

$$
E_{0b} = \frac{\sigma_0}{\sqrt{3}} \iint_{S_d} |[u_\tau]|_{0b} dS = \frac{2B\sigma_0}{\sqrt{3}} \int_0^H |[u_\tau]|_{0b} \sqrt{1 + \left(\frac{dy}{dx}\right)^2} \, dx
$$

$$
= \frac{2BH\sigma_0}{\sqrt{3}} \int_0^{\pi/2} |[u_\tau]|_{0b} \sqrt{1 + \frac{1}{h^2\cos^2\vartheta}\left(\frac{d\varsigma}{d\vartheta}\right)^2} \cos\vartheta d\vartheta.
$$

(5.61)

The derivative $d\varsigma/d\vartheta$ should be found along the curve $0b$. Therefore, it can be excluded by means of Eq. (5.52) and Eq. (5.61) becomes

$$\frac{E_{0b}}{BW^2\omega\sigma_0}$$

$$= \frac{2h}{\sqrt{3}} \int_0^{\pi/2} \left\{ (1 - \sin\vartheta)\varsigma_{0b}(\vartheta) + \frac{\left[\dfrac{\varsigma_{0b}^2(\vartheta)}{2h} - (\beta_0 + \beta_1\cos\vartheta + \sin\vartheta)h\right]^2}{\varsigma_{0b}(\vartheta)(1 - \sin\vartheta)} \right\} \cos\vartheta d\vartheta.$$

$$(5.62)$$

where $\varsigma_{0b}(\vartheta)$ should be replaced with the function of ϑ according to Eq. (5.56) and β_0 should be excluded by means of Eq. (5.55).

The amount of velocity jump across the velocity discontinuity line bd is simply equal to $|\mathbf{U}\cdot\mathbf{j} - \mathbf{u}\cdot\mathbf{j}|$ at $\vartheta = \pi/2$. Therefore, using Eqs. (5.48), (5.49) and (5.55) leads to

$$|[u_\tau]|_{bd} = \frac{\omega H}{2}\left(\frac{\varsigma^2}{h^2} + \frac{\pi\beta_1}{2} - 1\right). \tag{5.63}$$

It has been assumed here that

$$2\varsigma^2 + h^2(\pi\beta_1 - 2) \geq 0 \tag{5.64}$$

in the range $\varsigma_b \leq \varsigma \leq 1$. Since the right hand side of Eq. (5.64) is a monotonically increasing function of ς, it is sufficient to verify the inequality (5.64) at $\varsigma = \varsigma_b$. The latter immediately follows from Eq. (5.57). An infinitesimal length element of the velocity discontinuity line bd is dy or, with the use of Eq. (5.47), $Wd\varsigma$. Therefore, the rate of work dissipation at this line is obtained with the use of Eq. (5.63) as

$$E_{bd} = \frac{\sigma_0}{\sqrt{3}}\iint_{S_d} |[u_\tau]|_{bd}dS = \frac{BHW\omega\sigma_0}{\sqrt{3}}\int_{\varsigma_b}^1 \left(\frac{\varsigma^2}{h^2} + \frac{\pi\beta_1}{2} - 1\right)d\varsigma.$$

Integrating yields

$$\frac{E_{bd}}{BW^2\omega\sigma_0} = \frac{h}{\sqrt{3}}\left[\frac{1 - \varsigma_b^3}{3h^2} + \left(\frac{\pi\beta_1}{2} - 1\right)(1 - \varsigma_b)\right]. \tag{5.65}$$

Equation (5.40) is valid. Therefore, using this equation the inequality (1.4) for the problem under consideration can be represented in the form

$$g_u = \frac{G_u}{2BW^2\sigma_0} = \frac{(E_V + E_{0b} + E_{bd})}{\omega BW^2\sigma_0}. \tag{5.66}$$

Substituting Eqs. (5.59), (5.62) and (5.65) into Eq. (5.66) gives the value of g_u which depends on β_1. The right hand side of Eq. (5.66) should be minimized with respect to β_1 to find the best upper bound bending moment based on the kinematically admissible velocity field chosen. This minimization along with

Fig. 5.7 Comparison of two solutions for the limit moment

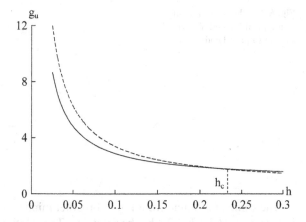

integration in Eqs. (5.59) and (5.62) should be carried out numerically. However, there are some difficulties with applying standard numerical techniques. First, the function $\varsigma_{0b}(\vartheta)$ involved in Eqs. (5.59) and (5.62) reduces to the expression $0/0$ at $\vartheta = \pi/2$. In order to overcome this difficulty, the right hand side of Eq. (5.56) can be expanded in a series in the vicinity of $\vartheta = \pi/2$. Then,

$$\varsigma_{0b}(\vartheta) = h\sqrt{1 - \frac{\beta_1 \pi}{2}} + \frac{2\beta_1 h}{3}\left(1 - \frac{\beta_1 \pi}{2}\right)^{-1/2}\left(\frac{\pi}{2} - \vartheta\right) + o\left(\frac{\pi}{2} - \vartheta\right), \qquad \vartheta \to \frac{\pi}{2}.$$

$$(5.67)$$

The first two terms can be used to replace the function $\varsigma_{0b}(\vartheta)$ in the range $\pi/2 - \delta \le \vartheta \le \pi/2$ where $\delta \ll 1$. Then, integration over this range can be performed analytically and integration over the range $0 \le \vartheta \le \pi/2 - \delta$ can be carried out numerically with no difficulty.

Another difficulty is that the integrand in Eq. (5.62) approaches infinity as $\vartheta \to 0$. It is possible to show that the integral is convergent. Expanding the integrand in a series in the vicinity of $\vartheta = 0$ Eq. (5.62) can be rewritten in the form

$$\frac{E_{0b}}{BW^2 \omega \sigma_0} = \frac{h^2 [2\beta_1 - (1 + \beta_1 \pi/2)]^2 \sqrt{\delta}}{\sqrt{3}\sqrt{1 - 2\beta_1(1 - \pi/4)}} + \frac{2h}{\sqrt{3}}I. \qquad (5.68)$$

where

$$I = \int\limits_{\delta}^{\pi/2} \left\{ (1 - \sin\vartheta)\varsigma_{0b}(\vartheta) + \frac{[\varsigma_{0b}^2(\vartheta) - 2(\beta_0 + \beta_1 \cos\vartheta + \sin\vartheta)h^2]^2}{4h^2 \varsigma_{0b}(\vartheta)(1 - \sin\vartheta)} \right\} \cos\vartheta d\vartheta.$$

Using Eq. (5.67) numerical integration here can be carried out with no difficulty. Then, E_{0b} is determined from Eq. (5.68).

Fig. 5.8 Schematic drawing
of a typical welded T-joint
subject to pure bending

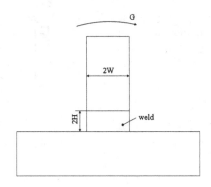

The value of g_u found from Eq. (5.66) after minimization is depicted in Fig. 5.7 in the range $0.03 \leq h \leq 0.4$ by the solid line. The solution found is valid if and only if $\varsigma_b \leq 1$ (i.e. the y-coordinate of point b must be less or equal to the y-coordinate of point d as it is shown in Fig. 5.6). This inequality has been verified using Eq. (5.57) and the value of β_1 found in course of numerical minimization. The broken line in Fig. 5.7 corresponds to the solution given by Eq. (5.42). It is seen that the lines intersect. The present solution provides a better (lower) value of the plastic limit moment as compared to the solution given by Eq. (5.42) where the solid line is under the broken line and *vice versa*. Equating the right hand sides of Eqs. (5.42) and (5.66) after minimization gives the equation for the value of $h = h_c$ at which both solutions result in the same plastic limit moment. The solution to this equation is $h_c \approx 0.22$. Thus, as it is seen from Fig. 5.7, the solution given by Eq. (5.42) should be used in the range $h \geq h_c$ and the solution given by Eq. (5.66) in the range $h \leq h_c$.

It is worthy of note that the solution found is also applicable to highly undermatched welded T-joints which are of great practical importance (Fig. 5.8).

References

S. Alexandrov, Limit load in bending of welded samples with a soft welded joint. J. Appl. Mech. Tech. Phys. **49**, 340–345 (2008)

S. Alexandrov, M. Kocak, Limit load analysis of strength undermatched welded T-joint under bending. Fatigue Fract. Eng. Mater. Struct. **30**, 351–355 (2007)

R. Hill, *The Mathematical Theory of Plasticity* (Clarendon Press, Oxford, 1950)

Y.-J. Kim, K.-H. Schwalbe, Mismatch effect on plastic yield loads in idealised weldments I. Weld centre cracks. Eng. Fract. Mech. **68**, 163–182 (2001a)

Y.-J. Kim, K.-H. Schwalbe, Mismatch effect on plastic yield loads in idealised weldments II. Heat affected zone cracks. Eng. Fract. Mech. **68**, 183–199 (2001b)

Y.-J. Kim, K.-H. Schwalbe, Compendium of yield load solutions for strength mis-matched DE(T), SE(B) and C(T) specimens. Eng. Fract. Mech. **68**, 1137–1151 (2001c)

Chapter 6
Miscellaneous Topics

This chapter deals with several typical solutions to demonstrate effects of the mis-match factor, three-dimensional deformation (thickness effect) and plastic anisotropy on the limit load of welded plates. The general approach to build up singular kinematically admissible velocity fields previously used in plane strain and axisymmetric solutions for highly undermatched structures is extended to three-dimensional deformation of tensile specimens to estimate the thickness effect on the limit load. The effect of the mis-match factor on the limit load is illustrated by simple examples for overmatched and undermatched center cracked specimens in tension. The effect of plastic anisotropy on the limit load is demonstrated for both highly undermatched and overmatched center cracked specimens under plane strain conditions.

6.1 Thickness Effect for Highly Undermatched Tensile Specimens

A typical tensile specimen with a crack is illustrated in Fig. 2.1. Its cross-section by the plane of symmetry containing the crack is shown in Fig. 6.1. Even though the solution given by Eq. (2.7) is a correct upper bound solution independently of the ratio $B/(W - a)$, the assumption that the principal normal strain rate in the thickness direction vanishes is reasonable if and only if this ratio is large enough. It is obvious that this condition is not satisfied in many cases. Therefore, it is important to estimate the effect of B-value on the limit load. To this end, it is necessary to use three-dimensional kinematically admissible velocity fields. Two solutions of this class are given below. One of these solutions is for center cracked specimens and the other is for double edge cracked specimens. The former solution is compared to the plane strain solution given in Chap. 2.

S. Alexandrov, *Upper Bound Limit Load Solutions for Welded Joints with Cracks*, SpringerBriefs in Computational Mechanics, DOI: 10.1007/978-3-642-29234-7_6, © The Author(s) 2012

Fig. 6.1 Cross-section of the tensile specimen by the plane of symmetry

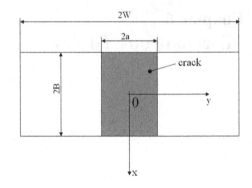

Fig. 6.2 Geometry of the specimen–notation

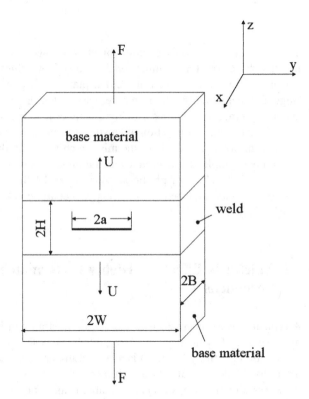

6.1.1 Center Cracked Specimen

The geometry of the specimen and the orientation of the axes of Cartesian coordinates (x, y, z) are shown in Fig. 6.2. Equation (1.13) is valid for this specimen. Therefore, it is sufficient to find a solution for the specimen with no crack. Such solutions have been proposed in Alexandrov (1999) and Alexandrov and Kocak (2008). The latter is provided below.

The specimen has three planes of symmetry, and the axes of the Cartesian coordinate system can be chosen along the three straight lines of intersection of these planes. Therefore, it is sufficient to get the solution in the domain $x \geq 0$, $y \geq 0$ and $z \geq 0$. By analogy to the plane strain solution (see Eqs. (2.9) and (2.11)), assume the following kinematically admissible velocity field

$$\frac{u_z}{U} = \eta, \quad \frac{u_x}{U} = \alpha v + f(\eta), \quad \frac{u_y}{U} = -(1 + \alpha)\mu + g(\eta). \tag{6.1}$$

Here $\eta = z/H$, $v = x/H$ and $\mu = y/H$ are the dimensionless coordinates, $f(\eta)$ and $g(\eta)$ are arbitrary functions of η, and α is an arbitrary constant. The plane strain velocity field given by Eqs. (2.9) and (2.11) is obtained as a special case at $\alpha = 0$ and $f(\eta) = 0$. The velocity field (6.1) satisfies the equation of incompressibility (1.6) at any $f(\eta)$, $g(\eta)$ and α. Using Eq. (1.9) and the symmetry conditions $\zeta_{xy} = 0$ and $\zeta_{yz} = 0$ at $\eta = 0$ as additional requirements on kinematically admissible velocity fields possible and most simple functions $f(\eta)$ and $g(\eta)$ are $f(\eta) = f_0 + f_1\sqrt{1 - \eta^2}$ and $g(\eta) = g_0 + g_1\sqrt{1 - \eta^2}$, where f_0, f_1, g_0 and g_1 are arbitrary constants. Then, the velocity field (6.1) becomes

$$\frac{u_z}{U} = \eta, \quad \frac{u_x}{U} = \alpha v + f_0 + f_1\sqrt{1 - \eta^2}, \quad \frac{u_y}{U} = -(1 + \alpha)\mu + g_0 + g_1\sqrt{1 - \eta^2}. \tag{6.2}$$

To satisfy the natural velocity boundary conditions $u_x = 0$ at $x = 0$ and $u_y = 0$ at $y = 0$, it is necessary to introduce a rigid zone in the vicinity of the planes of symmetry $x = 0$ and $y = 0$. This zone moves up along with the block of rigid base material whose velocity is U. The boundary of the rigid zone and the plastic zone is a piece-wise smooth surface consisting of two smooth parts. The structure of the velocity field (6.2) and the position of the axes of symmetry $x = 0$ and $y = 0$ require that the unit normal vectors to these smooth parts are represented by the equations

$$\mathbf{n} = -\sin\varphi\mathbf{j} + \cos\varphi\mathbf{k}, \quad \mathbf{m} = -\sin\phi\mathbf{i} + \cos\phi\mathbf{k} \tag{6.3}$$

where $\mathbf{i}$, $\mathbf{j}$ and $\mathbf{k}$ are the unit vectors parallel to x, y, and z axes, respectively, φ is the orientation of the line (in planes $x = $ constant) tangent to the velocity discontinuity surface (curve in planes $x = $ constant) relative to y axis, and ϕ is the orientation of the line (in planes $y = $ constant) tangent to the velocity discontinuity surface (curve in planes $y = $ constant) relative to x axis. The cross-section of the velocity discontinuity surface corresponding to the unit vector $\mathbf{n}$ by a plane $x = $ constant and angle φ are illustrated in Fig. 6.3 (ψ introduced in this figure will be defined later). It follows from this figure and the definition for η and μ that

$$\tan\varphi = \frac{dz}{dy} = \frac{d\eta}{d\mu}. \tag{6.4}$$

The velocity vector in the rigid zone can be written as

Fig. 6.3 Shape of velocity
discontinuity surface at
$x =$ constant

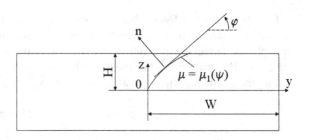

$$\mathbf{U} = U\mathbf{k}. \tag{6.5}$$

The velocity vector in the plastic zone is represented by

$$\mathbf{u} = u_x\mathbf{i} + u_y\mathbf{j} + u_z\mathbf{k} \tag{6.6}$$

where u_x, u_y and u_z are given by Eq. (6.2). Assume that $\mathbf{U} \equiv \mathbf{u}_1$ and $\mathbf{u} \equiv \mathbf{u}_2$ in Eq. (1.7). Then, it follows from Eqs. (6.3), (6.4), (6.5), and (6.6) that

$$\frac{d\mu}{d\psi} = \frac{[2(1 + \alpha)\mu - g_0 - g_1 \cos 2\psi] \cos 2\psi}{1 - \sin 2\psi} \tag{6.7}$$

where

$$\sin 2\psi = \eta, \quad d\eta = 2 \cos 2\psi d\psi. \tag{6.8}$$

Equation (6.7) determines the same curve $\mu = \mu_1(\psi)$ in each yz plane (Fig. 6.3). This curve must contain the point $z = 0$ and $y = 0$. Therefore, the boundary condition to Eq. (6.7) is

$$\mu = 0 \tag{6.9}$$

at $\eta = 0$ (or $\psi = 0$). A natural additional requirement is that the curve has a common point with the line $\eta = 1$ (or $\psi = \pi/4$). Since the denominator of the right hand side of Eq. (6.7) vanishes at $\psi = \pi/4$, a necessary condition is that its nominator also vanishes at $\psi = \pi/4$. The latter condition is satisfied at the point

$$\mu = \mu_0 = \frac{g_0}{1 + \alpha} \tag{6.10}$$

Expanding the nominator and denominator of the right hand side of Eq. (6.7) in a series in the vicinity of $\psi = \pi/4$ and $\mu = \mu_0$ gives

$$\frac{d(\mu - \mu_0)}{d(\psi - \pi/4)} = -2(1 + \alpha)\frac{(\mu - \mu_0)}{(\psi - \pi/4)} - 4g_1.$$

At $-1 - \alpha \neq 1/2$, the solution to this equation is

$$\mu = \mu_0 - \frac{4g_1}{(3 + 2\alpha)}\left(\psi - \frac{\pi}{4}\right) + C_1\left(\frac{\pi}{4} - \psi\right)^{-2(1+\alpha)} \tag{6.11}$$

where C_1 is a constant of integration. It is expected to assume that the normal strain rates ζ_{xx} and ζ_{yy} are compressive, $\zeta_{xx} < 0$ and $\zeta_{yy} < 0$. Then, it follows from the velocity field (6.2) that $0 > \alpha > -1$. In this case, the condition $\mu = \mu_0$ at $\psi = \pi/4$ is satisfied if and only if $C_1 = 0$. Finally, the solution to Eq. (6.7) in the vicinity of $\psi = \pi/4$ is determined from Eqs. (6.10) and (6.11)

$$\mu = \mu_1(\psi) = \frac{g_0}{1+\alpha} - \frac{4g_1}{(3+2\alpha)}\left(\psi - \frac{\pi}{4}\right). \tag{6.12}$$

It is assumed that this equation is valid in the range

$$\pi/4 \geq \psi \geq \psi_1 = \pi/4 - \delta \tag{6.13}$$

where $\delta \ll 1$. Equation (6.7) is a linear differential equation of first order with respect to μ. Therefore, its general solution can be found with no difficulty. The particular solution satisfying the boundary condition (6.9) is

$$\mu_1(\psi) = -\frac{g_0}{(1+\alpha)}\left[1 - (1 - \sin 2\psi)^{-(1+\alpha)}\right] - 2g_1 \Phi_1(\psi)(1 - \sin 2\psi)^{-(1+\alpha)},$$
$$\Phi_1(\psi) = \int_0^\psi \cos^2 2\gamma(1 - \sin 2\gamma)^\alpha d\gamma. \tag{6.14}$$

Here γ is a dummy variable. The function $\mu_1(\psi)$ must be continuous in the range $0 \leq \psi \leq \pi/4$. Therefore, equating $\mu_1(\psi)$ found from Eqs. (6.12) and (6.14) at $\psi = \psi_1$ (ψ_1 has been introduced in Eq. (6.13)) determines the value of g_1 involved in Eq. (6.12)

$$g_1 = -\frac{g_0}{(1+\alpha)} \frac{(3+2\alpha)}{\left[4\delta(1 - \cos 2\delta)^{(1+\alpha)} + 2(3+2\alpha)\Phi_1(\psi_1)\right]}. \tag{6.15}$$

Equations (6.12), (6.14) and (6.15) determine the shape of the velocity discontinuity surface shown in Fig. 6.3 depending on one parameter, g_0. Substituting Eqs. (6.5) and (6.6) into Eq. (1.8) and using Eq. (6.8) give the amount of velocity jump across this velocity discontinuity surface denoted by $|[u_\tau]_1|$

$$|[u_\tau]_1| = U\sqrt{(1 - \sin 2\psi)^2 + (\alpha v + f_0 + f_1 \cos 2\psi)^2 + [(1+\alpha)\mu_1(\psi) - g_0 - g_1 \cos 2\psi]^2}. \tag{6.16}$$

Using Eq. (6.8) the infinitesimal area element of the velocity discontinuity surface can be found in the form

$$dS = dx\sqrt{(dy)^2 + (dz)^2} = H^2 dv\sqrt{(d\mu)^2 + (d\eta)^2}$$
$$= H^2 dv\sqrt{(d\mu)^2 + 4\cos^2 2\psi(d\psi)^2} . \tag{6.17}$$

Substituting the derivative $d\mu/d\psi$ from Eq. (6.7) into Eq. (6.17) leads to

$$dS = \frac{2H^2 v \cos 2\psi}{(1 - \sin 2\psi)} \sqrt{(1 - \sin 2\psi)^2 + [(1 + \alpha)\mu_1(\psi) - g_0 - g_1 \cos 2\psi]^2} d\psi dv.$$

(6.18)

The rate of work dissipation at the velocity discontinuity surface $\mu = \mu_1(\psi)$ is finally given by

$$E_1 = \frac{2\sigma_0 UH^2}{\sqrt{3}} \int\limits_0^{\pi/4} \int\limits_{v_1(\psi)}^{B/H} \frac{|[u_\tau]_1|}{U} \frac{\cos 2\psi}{(1 - \sin 2\psi)} \times \frac{1}{\sqrt{(1 - \sin 2\psi)^2 + [(1 + \alpha)\mu_1(\psi) - g_0 - g_1 \cos 2\psi]^2}} dv d\psi.$$

(6.19)

The function $v_1(\psi)$ will be determined later.

A similar analysis can be carried out for the velocity discontinuity surface corresponding to the unit normal vector $\mathbf{m}$ introduced in Eq. (6.3). As a result, the equation for this surface is $v = v_1(\psi)$ where

$$v_1(\psi) = -\frac{f_0}{\alpha} - \frac{4f_1}{(1 - 2\alpha)}\left(\psi - \frac{\pi}{4}\right)$$

(6.20)

in the interval $\pi/4 \geq \psi \geq \psi_1$ and

$$v_1(\psi) = -\frac{f_0}{\alpha} + (1 + \sin 2\psi)^\alpha\left[\frac{f_0}{\alpha} - 2f_1 \Phi_2(\psi)\right],$$

$$\Phi_2(\psi) = \int\limits_0^\psi \frac{\cos^2 \gamma}{(1 - \sin 2\gamma)^{(1+\alpha)}} d\gamma$$

(6.21)

in the interval $\psi_1 \geq \psi \geq 0$. Constant f_1 should be excluded and expressed as

$$f_1 = \frac{f_0(1 - 2\alpha)(1 - \cos 2\delta)^\alpha}{\alpha[4\delta + 2(1 - 2\alpha)(1 - \cos 2\delta)^\alpha \Phi_2(\psi_1)]}.$$

(6.22)

Thus the function $v_1(\psi)$ involved in Eq. (6.19) is determined by Eqs. (6.20) and (6.21). In particular, $v_1(\pi/4) = v_0 = -f_0/\alpha$. The rate of work dissipation at the velocity discontinuity surface $v = v_1(\psi)$ is given by

$$E_2 = \frac{2\sigma_0 UH^2}{\sqrt{3}} \int\limits_0^{\pi/4} \int\limits_{\mu_1(\psi)}^{W/H} \frac{|[u_\tau]_2|}{U} \frac{\cos 2\psi}{(1 - \sin 2\psi)} \times \frac{1}{\sqrt{(1 - \sin 2\psi)^2 + [\alpha v_1(\psi) + f_0 + f_1 \cos 2\psi]^2}} d\mu d\psi$$

(6.23)

Fig. 6.4 Configuration of plastic and rigid zones at $z = H$

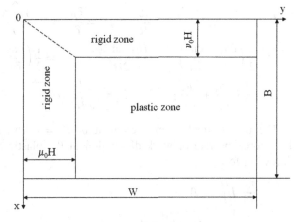

where

$$\left|[u_\tau]_2\right| = U\sqrt{(1 - \sin 2\psi)^2 + [\alpha v_1(\psi) + f_0 + f_1 \cos 2\psi]^2 + [(1 + \alpha)\mu - g_0 - g_1 \cos 2\psi]^2}.$$

$$(6.24)$$

It is worthy of note here that Eqs. (6.20) and (6.21) for the velocity discontinuity surface $v = v_1(\psi)$ depend on the single free parameter f_0 due to Eq. (6.22). The geometry of the rigid and plastic zones at $\eta = 1$ is shown in Fig. 6.4.

Another velocity discontinuity surface appears at $\eta = 1$ (or $\psi = \pi/4$) in the region $\mu_1(\pi/4) \le \mu \le W/H$ and $v_1(\pi/4) \le v \le B/H$ or, using Eqs. (6.12) and (6.20), $g_0/(1 + \alpha) \le \mu \le W/H$ and $-f_0/\alpha \le v \le B/H$. The amount of velocity jump across this velocity discontinuity surface is equal to $\left|[u_\tau]_3\right| = \sqrt{u_x^2 + u_y^2}$ where u_x and u_y should be taken at $\eta = 1$. Then, it follows from Eq. (6.2) that

$$\left|[u_\tau]_3\right| = U\sqrt{(f_0 + \alpha v)^2 + [g_0 - (1 + \alpha)\mu]^2}. \qquad (6.25)$$

The infinitesimal area element is just $dS = H^2 d\mu dv$. Therefore, using Eq. (6.25), the rate of work dissipation at this velocity discontinuity surface is given by

$$E_3 = \frac{\sigma_0 U H^2}{\sqrt{3}} \int\limits_{-f_0/\alpha}^{B/H} \int\limits_{g_0/(1+\alpha)}^{W/H} \sqrt{(f_0 + \alpha v)^2 + [g_0 - (1 + \alpha)\mu]^2} d\mu dv. \qquad (6.26)$$

Using Eqs. (1.3), (6.2) and (6.8) the components of the strain rate tensor and the equivalent strain rate are determined by

$$\zeta_{xx} = \frac{\partial u_x}{\partial x} = \frac{\alpha U}{H}, \quad \zeta_{yy} = \frac{\partial u_y}{\partial y} = -\frac{(1+\alpha)U}{H}, \quad \zeta_{zz} = \frac{\partial u_z}{\partial z} = \frac{U}{H},$$

$$\zeta_{xz} = \frac{1}{2}\left(\frac{\partial u_x}{\partial z} + \frac{\partial u_z}{\partial x}\right) = -\frac{U f_1 \tan 2\psi}{2H}, \quad \zeta_{yz} = \frac{1}{2}\left(\frac{\partial u_y}{\partial z} + \frac{\partial u_z}{\partial y}\right) = -\frac{U g_1 \tan 2\psi}{2H},$$

$$\zeta_{xy} = \frac{1}{2}\left(\frac{\partial u_x}{\partial y} + \frac{\partial u_y}{\partial x}\right) = 0, \quad \zeta_{eq} = \frac{U}{\sqrt{3}H}\sqrt{4(1+\alpha+\alpha^2) + (f_1^2 + g_1^2)\tan^2 2\psi}.$$

The infinitesimal volume element is $dV = H^3 dv d\mu d\eta = 2H^3 \cos 2\psi dv d\mu d\psi$. Therefore, the rate of work dissipation in the plastic zone is determined in the following form

$$E_4 = \frac{\sigma_0}{\sqrt{3}} \iiint_V \zeta_{eq} dV$$

$$= \frac{2U\sigma_0 H^2}{\sqrt{3}} \int_0^{\pi/4} \int_{\mu_1(\psi)}^{W/H} \int_{v_1(\psi)}^{B/H} \sqrt{4(1+\alpha+\alpha^2) + (f_1^2 + g_1^2)\tan^2 2\psi} \cos 2\psi dv d\mu d\psi.$$

Integrating with respect to v and μ gives

$$E_4 = \frac{2U\sigma_0 H^2}{\sqrt{3}} \int_0^{\pi/4} \frac{\left(\frac{B}{H} - v_1(\psi)\right)\left(\frac{W}{H} - \mu_1(\psi)\right)}{\times \sqrt{4(1+\alpha+\alpha^2) + (f_1^2 + g_1^2)\tan^2 2\psi}} \cos 2\psi d\psi. \tag{6.27}$$

The rate at which external forces F do work is (two equal forces act and one-eighth of the specimen is considered)

$$\iint_{S_v} (t_i v_i) dS = \frac{1}{4} FU. \tag{6.28}$$

The rate of work dissipation in one-eighth of the specimen is $E_1 + E_2 + E_3 + E_4$. Therefore, it follows from Eqs. (1.4) and (6.28) that the dimensionless upper bound limit load is determined by

$$f_u^{(1)} = \frac{F_u^{(1)}}{4BW\sigma_0} = \frac{E_1 + E_2 + E_3 + E_4}{UBW\sigma_0}. \tag{6.29}$$

The right hand side of this equation can be found by means of Eqs. (6.19), (6.23), (6.26), and (6.27), and using Eqs. (6.12), (6.14), (6.15), (6.16), (6.20), (6.21), (6.22), and (6.24), can be represented as a function of three parameters, α, f_0 and g_0. Then, the right hand side of Eq. (6.29) should be minimized with respect to these three parameters to find the best upper bound limit load based on the kinematically admissible velocity field chosen. The minimization has been performed numerically in the domain $2 \leq W/H \leq 10$ and $2 \leq B/H \leq 10$ It is worthy of note that W and B are involved in Eq. (6.29) in a symmetric manner.

Fig. 6.5 Kinematically
admissible velocity field
for specimens with small
ratio W/H

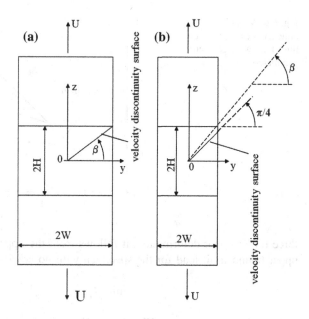

Therefore, $f_u^{(1)}$ is an even function of $W - B$. Using this property, the numerical solution can be fitted to a polynomial as

$$f_u^{(1)} = q_0 + q_1 \frac{(W-B)^2}{H^2} + q_2 \frac{(W+B)}{H} + q_3 \frac{(W-B)^2(W+B)}{H^3} + q_4 \frac{(W+B)^2}{H^2}$$
(6.30)

where $q_0 \approx 1.122$, $q_1 \approx -0.0264$, $q_2 \approx 0.1035$, $q_3 \approx 0.00095$, and $q_4 \approx 0.00025$.

For the specimen with a sufficiently small ratio W/H which is equivalent to a sufficiently large crack for the specimen with the crack, another solution used in many previous studies (see, for example, Kotousov and Jaffar (2006)) can be proposed by assuming the kinematically admissible velocity field consisting of two rigid blocks (in the domain $y \geq 0$ and $z \geq 0$) separated by a velocity discontinuity surface (straight line in yz planes), as shown in Fig. 6.5. After some simple algebra, the upper bound on the limit load based on this special velocity field can be found as

$$f_u^{(2)} = \frac{F_u^{(2)}}{4BW\sigma_0} = \begin{cases} \dfrac{2}{\sqrt{3}}, & \text{if } \beta \geq \dfrac{\pi}{4} \\[2ex] \dfrac{2}{\sqrt{3}\sin 2\beta}, & \text{if } \beta < \dfrac{\pi}{4} \end{cases}$$
(6.31)

where $\beta = \arctan(H/W)$. The velocity field corresponding to $\beta < \pi/4$ is shown in Fig. 6.5a and that to $\beta \geq \pi/4$ in Fig. 6.5b. Since the solution (2.7) at $a = 0$ and the solution (6.31) are based on kinematically admissible velocity fields applicable for

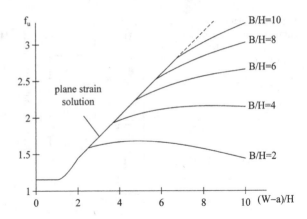

Fig. 6.6 Variation of the dimensionless upper bound limit load with geometric parameters

three dimensional deformation, it follows from the upper bound theorem that the upper bound limit load for the specimen with no crack is

$$f_u = \min\left\{f_u^{(1)}, f_u^{(2)}, f_u^{(3)}\right\}. \tag{6.32}$$

Here $f_u^{(3)}$ denotes $F_u/(4BW\sigma_0)$ from Eq. (2.7). In order to find the limit load for the specimen with a crack, it is just necessary to adopt Eq. (1.13).

The variation of the dimensionless upper bound limit load with $(W-a)/H$ at different values of B/H is depicted in Fig. 6.6. This diagram demonstrates that the error introduced by the assumption of plain strain can be quite significant. In particular, the single curve (including its extension shown by a dashed line) independent of B/H corresponds to the plane-strain solution given in Eq. (2.7) whereas five different curves corresponding to five different values of B/H demonstrate the effect of three dimensional deformation on the limit load. Note that the actual effect of three dimensional deformation is even more significant than that shown in Fig. 6.6. For, the solution (2.7) is an accurate approximation of a numerical slip-line solution satisfying all field equations and boundary conditions, whereas the limit load solution given in Eq. (6.30) is based on the minimization of a simple function of three variables. The exact three dimensional solution would result in a smaller limit load and therefore shift the curves corresponding to three dimensional deformation down.

6.1.2 Double Edge Cracked Specimen

The geometry of the specimen and the orientation of the axes of Cartesian coordinates (x, y, z) are shown in Fig. 6.7. There are two cracks of length a. The specimen is loaded by two equal forces F whose magnitude at plastic collapse should be evaluated. The blocks of rigid base material move with a velocity U along the z-axis in the opposite directions. The specimen has three planes of

Fig. 6.7 Geometry of the specimen–notation

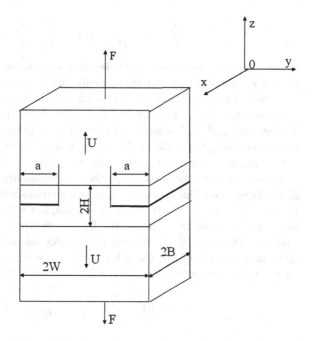

symmetry. The lines of intersection of each pair of these planes are parallel to the axes of the Cartesian coordinate system. It is assumed that the origin of this coordinate system is located at the intersection of the three planes of symmetry of the specimen. Because of the symmetry, it is sufficient to find the solution in the domain $x \geq 0$, $y \geq 0$ and $z \geq 0$. Let u_x, u_y and u_z be the velocity components in the directions of x, y and z, respectively. The velocity boundary conditions are

$$u_z = 0 \tag{6.33}$$

at $z = 0$ in the range $0 \leq y \leq W - a$,

$$u_z = U \tag{6.34}$$

at $z = H$,

$$u_x = 0 \tag{6.35}$$

at $x = 0$, and

$$u_y = 0 \tag{6.36}$$

at $y = 0$. A three-dimensional solution to this problem has been proposed by Alexandrov and Hwang (2010) and is given below.

Assume that $u_y = 0$ everywhere. Then, the boundary condition (6.36) is automatically satisfied. In planes xz the velocity field found in Hill (1950) for compression of a plastic layer between parallel plates can be adopted. In particular,

$$\frac{u_x}{U} = C - \frac{x}{H} - 2\sqrt{1 - \frac{z^2}{H^2}}, \quad \frac{u_z}{U} = \frac{z}{H} \tag{6.37}$$

where C is an arbitrary constant whose value will be determined later. It is possible to verify by inspection that the velocity field (6.37) satisfies the equation of incompressibility (1.6) and the asymptotic singular behaviour (1.9) in the vicinity of the surface $z = H$. The boundary conditions (6.33) and (6.34) are also satisfied. The velocity field (6.37) cannot be extended to the plane of symmetry $x = 0$ where $u_x = 0$ according to Eq. (6.35). Therefore, it is necessary to introduce a rigid zone in the vicinity of the plane $x = 0$. This zone moves along with the block of rigid base material along the z-axis. Therefore, the boundary condition (6.35) is automatically satisfied. One quarter of an arbitrary cross-section of the weld material at $y = y_0$, $\;0 \le y_0 < W - a$, is shown in Fig. 6.8. Let $\mathbf{n}$ be the unit normal vector to the velocity discontinuity surface $0b$. By assumption, its projection on the y-axis vanishes. Therefore, it is sufficient to consider the velocity discontinuity curve $0b$ in an arbitrary xz-plane. Using Eq. (6.37) the vector velocity field in the plastic zone can be represented as

$$\mathbf{u} = U\left(C - \frac{x}{H} - 2\sqrt{1 - \frac{z^2}{H^2}}\right)\mathbf{i} + U\frac{z}{H}\mathbf{k} \tag{6.38}$$

where $\mathbf{i}$ and $\mathbf{k}$ are the unit vectors along the x- and z- axes, respectively. The velocity vector in the rigid zone is

$$\mathbf{U} = U\mathbf{k}. \tag{6.39}$$

The unit normal vector $\mathbf{n}$ can be represented in the form (Fig. 6.8)

$$\mathbf{n} = -\sin\varphi\mathbf{i} + \cos\varphi\mathbf{k} \tag{6.40}$$

where φ is the angle between the tangent to the velocity discontinuity curve and the x-axis, measured clockwise from the axis. It is convenient to introduce the following dimensionless quantities

$$\varsigma = \frac{x}{B}, \quad \eta = \frac{z}{H}, \quad b = \frac{B}{H}, \quad \alpha = \frac{a}{W}, \quad \psi = \frac{W - a}{H}. \tag{6.41}$$

Assume that $\mathbf{U} \equiv \mathbf{u_1}$ and $\mathbf{u} \equiv \mathbf{u_2}$ in Eq. (1.7). Then, it follows from Eqs. (6.38), (6.39) and (6.40) that $\left(b\varsigma + 2\sqrt{1 - \eta^2} - C\right)\sin\varphi + (\eta - 1)\cos\varphi = 0$. By definition, $\tan\varphi = dz/dx$ (Fig. 6.8). Using Eq. (6.41) this equation can be transformed to

$$\frac{d\varsigma}{d\eta} = \frac{\left(b\varsigma + 2\sqrt{1 - \eta^2} - C\right)}{b(1 - \eta)}. \tag{6.42}$$

The solution to this equation determines the shape of the velocity discontinuity curve $0b$ (Fig. 6.8). The velocity field used is kinematically admissible if and only if this curve contains the origin of the coordinate system. Therefore, the boundary

Fig. 6.8 General structure of the kinematically admissible velocity field in xz planes

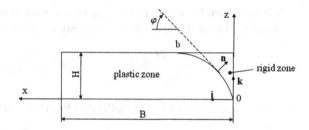

condition to Eq. (6.42) is $\varsigma = 0$ for $\eta = 0$. Equation (6.42) is a linear differential equation of first order. Therefore, its general solution can be found with no difficulty. Then, the particular solution satisfying the boundary condition formulated can be obtained as

$$\varsigma b = \frac{\eta\sqrt{1 - \eta^2} + \arcsin \eta - C\eta}{1 - \eta}. \tag{6.43}$$

The denominator of the right hand side of this equation vanishes at $\eta = 1$. Therefore, the velocity discontinuity curve $0b$ may have a common point with the line $\eta = 1$ (or $z = H$) if and only if the numerator vanishes at $\eta = 1$. This condition results in $C = \pi/2$. In this case it follows from Eq. (6.43) that

$$\varsigma_b = \lim_{\eta \to 1} \varsigma(\eta) = \frac{\pi}{2b}. \tag{6.44}$$

Thus the x-coordinate of point b (Fig. 6.8) is $x_b = \varsigma_b B = \pi H/2$ and the solution is valid if and only if $x_b \le B$ or $b \ge \pi/2$. Using Eq. (6.43) and the value of C found the equation for the velocity discontinuity curve $0b$ can be written as

$$\varsigma = \varsigma_{0b}(\eta) = \frac{\eta\sqrt{1 - \eta^2} + \arcsin \eta - \pi\eta/2}{b(1 - \eta)}. \tag{6.45}$$

The amount of velocity jump across this curve is determined from Eqs. (1.8), (6.38) and (6.39) as

$$\|[u_\tau]\|_{0b} = \sqrt{u_x^2 + (u_z - U)^2}. \tag{6.46}$$

Substituting Eq. (6.37) at $C = \pi/2$ into Eq. (6.46) and taking into account Eq. (6.41) lead to

$$\|[u_\tau]\|_{0b} = U\sqrt{\left(\frac{\pi}{2} - \varsigma b - 2\sqrt{1 - \eta^2}\right)^2 + (\eta - 1)^2}. \tag{6.47}$$

The infinitesimal length element of the velocity discontinuity curve $0b$ is determined with the use of Eqs. (6.41) and (6.42) at $C = \pi/2$ in the form

$$dl_{0b} = \sqrt{dx^2 + dz^2} = \frac{H}{(1 - \eta)}\sqrt{(1 - \eta)^2 + \left(\varsigma b + 2\sqrt{1 - \eta^2} - \frac{\pi}{2}\right)^2}\, d\eta. \tag{6.48}$$

It is seen from Eq. (6.47) that $\|[u_\tau]\|_{0b}$ is independent of y. Therefore, the rate of work dissipation at the velocity discontinuity curve $0b$ is expressed as

$$E_{0b} = \frac{\sigma_0}{\sqrt{3}} \iint\limits_{S_d} \|[u_\tau]\|_{0b} dS = \frac{\sigma_0}{\sqrt{3}} \left(1 - \frac{a}{W}\right) \int \|[u_\tau]\|_{0b} dl_{0b}. \tag{6.49}$$

Substituting Eqs. (6.47) and (6.48) into Eq. (6.49) and using Eq. (6.41) give

$$\frac{E_{0b}}{UBW\sigma_0} = \frac{(1-\alpha)}{\sqrt{3}b} \int\limits_0^1 \left[\left(\frac{\pi}{2} - \varsigma b - 2\sqrt{1-\eta^2}\right)^2 + (1-\eta)^2\right](1-\eta)^{-1} d\eta. \tag{6.50}$$

Replacing here ς with $\varsigma_{0b}(\eta)$ by means of Eq. (6.45) the integral can be evaluated numerically with no difficulty.

Another velocity discontinuity surface occurs at $y = W - a$. Its area is equal to the area of the plastic zone shown in Fig. 6.8. Since the rigid block moves along the z-axis with velocity U, the amount of velocity jump across this surface is given by Eq. (6.47). Also, the infinitesimal area element is simply $dS = dxdz$ or, with the use of Eq. (6.41), $dS = BHd\varsigma d\eta$. Therefore, the rate of work dissipation at this velocity discontinuity surface, E_1, is

$$\frac{E_1}{UBW\sigma_0} = \frac{1}{\sqrt{3}UWB} \iint\limits_{S_d} \|[u_\tau]\|_{0b} dS$$

$$= \frac{(1-\alpha)}{\sqrt{3}\psi} \int\limits_0^1 \int\limits_{\varsigma_{0b}(\eta)}^1 \sqrt{\left(\frac{\pi}{2} - \varsigma b - 2\sqrt{1-\eta^2}\right)^2 + (1-\eta)^2} d\varsigma d\eta. \tag{6.51}$$

Replacing $\varsigma_{0b}(\eta)$ in the lower limit of integration with the function of η by means of Eq. (6.45) the integral here can be evaluated numerically with no difficulty.

Still another velocity discontinuity surface occurs at $z = H$ in the range $\varsigma_b \leq \varsigma \leq 1$. The amount of velocity jump across this surface is just $|u_x|$ where the value of u_x should be found from Eq. (6.37) at $z = H$ and $C = \pi/2$. The infinitesimal area element of this velocity discontinuity surface is $dS = dxdy$. Therefore, the rate of work dissipation, E_2, is determined as

$$\frac{E_2}{UBW\sigma_0} = \frac{1}{\sqrt{3}UBW} \iint |u_x| dxdy. \tag{6.52}$$

Substituting Eq. (6.37) at $C = \pi/2$ and $z = H$ into Eq. (6.52), taking into account Eqs. (6.41) and (6.44), and integrating give

$$\frac{E_2}{UBW\sigma_0} = \frac{(1-\alpha)}{\sqrt{3}} \int\limits_{\varsigma_b}^1 \left|\frac{\pi}{2} - \varsigma b\right| d\varsigma = \frac{(1-\alpha)(b - \pi/2)^2}{2\sqrt{3}b}. \tag{6.53}$$

It has been taken into account here that $\varsigma b - \pi/2 \geq 0$ in the range $\varsigma_b \leq \varsigma \leq 1$ due to Eq. (6.44). Using Eqs. (1.3), (6.37) and (6.41) the non-zero strain rate components and the equivalent strain rate can be determined as

$$\zeta_{xx} = \frac{\partial u_x}{\partial x} = -\frac{U}{H}, \quad \zeta_{zz} = \frac{\partial u_z}{\partial z} = \frac{U}{H}, \quad \zeta_{xz} = \frac{1}{2}\left(\frac{\partial u_x}{\partial z} + \frac{\partial u_z}{\partial x}\right) = \frac{U}{H}\frac{\eta}{\sqrt{1-\eta^2}},$$

$$\zeta_{eq} = \frac{2U}{\sqrt{3}H\sqrt{1-\eta^2}}.$$

$$(6.54)$$

Since $dV = dxdydz$ and ζ_{eq} is independent of both ς and y, the rate of work dissipation in the plastic zone is

$$E_V = \sigma_0 \iiint\limits_V \zeta_{eq}dV = \sigma_0 UBW \left[\frac{\pi(1-\alpha)}{\sqrt{3}}\left(1 - \frac{\pi}{2b}\right) + \frac{2(1-\alpha)}{\sqrt{3}}\int\limits_0^1 \frac{\varsigma_{0b}(\eta)}{\sqrt{1-\eta^2}}d\eta\right].$$

$$(6.55)$$

Here Eq. (6.41) has been taken into account. Replacing $\varsigma_{0b}(\eta)$ in the integrand with the function of η by means of Eq. (6.45) the integral can be evaluated numerically with no difficulty.

The rate at which external forces F do work is (two equal forces act and one-eighth of the specimen is considered)

$$\iint\limits_{S_v} (t_i v_i)dS = \frac{1}{4}FU. \tag{6.56}$$

The rate of work dissipation in one-eighth of the specimen is $E_1 + E_2 + E_{0b} + E_V$. Therefore, it follows from Eqs. (1.4) and (6.56) that

$$f_u^{(1)} = \frac{F_u^{(1)}}{4BW\sigma_0} = \frac{E_1 + E_2 + E_{0b} + E_V}{UBW\sigma_0}. \tag{6.57}$$

The right hand side of this equation can be found by means of Eqs. (6.50), (6.51), (6.53), and (6.55).

For specimens with small cracks the contribution of E_1 to the rate of total work dissipation is too large. Therefore, a better solution can be obtained by ignoring the existence of the crack. To this end, it is just necessary to put $\alpha = 0$ and to exclude E_1 in Eq. (6.57). As a result, the value of $f_u^{(2)} = F_u^{(2)}/(4BW\sigma_0)$ can be determined. Having the values of $f_u^{(1)}$ and $f_u^{(2)}$ the upper bound limit load should be found as

$$f_u = \min\left\{f_u^{(1)}, f_u^{(2)}\right\}. \tag{6.58}$$

It has been mentioned after Eq. (6.44) that the solution found is valid only if $b \geq \pi/2$. However, it is straightforward to extend it to the range $b < \pi/2$. In this

Fig. 6.9 Variation of η_1 with b

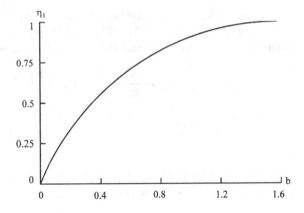

case the curve $0b$ (Fig. 6.8) intersects the line $x = B$ (or $\varsigma = 1$). Using Eq. (6.45) the corresponding value of $\eta = \eta_1$ can be determined from the following equation

$$\frac{\eta_1\sqrt{1 - \eta_1^2} + \arcsin\eta_1 - \pi\eta_1/2}{b(1 - \eta_1)} = 1.$$

This equation should be solved numerically in the range $\pi/2 > b > 0$ and its solution is illustrated in Fig. 6.9. The corresponding limit load can be obtained from Eq. (6.57) putting $E_2 = 0$. Moreover, Eq. (6.55) becomes

$$E_V = \frac{2\sigma_0 UBW(1 - \alpha)}{\sqrt{3}} \int\limits_0^{\eta_1} \frac{\varsigma_{0b}(\eta)}{\sqrt{1 - \eta^2}} d\eta$$

and the upper limit of integration with respect to η in Eq. (6.51) should be changed to η_1. The equation for the limit load in the case of small cracks should be treated in a similar manner.

In order to show the significance of three-dimensional solutions for a class of specimens, comparison between the present solution and the plane strain solution reported in Kim and Schwalbe (2001) is made. It is obvious that the difference between the solutions mainly depends on the value of B. Therefore, it is reasonable to compare the limit load for several values of B keeping a constant value of α, say $\alpha = 0.2$. The solution in Kim and Schwalbe (2001) provides the variation of the limit load with ψ in the form of an approximating function. Since the value of α is fixed, the value of ψ varies because the value of H changes (see Eq. (6.41)). Therefore, the value of b also changes with ψ if the value of B is kept a constant. The variation of the limit load found from Eq. (6.58) is depicted in Fig. 6.10 for three cases, namely $\psi = b$ (or $W - a = B$), $\psi = 2b$ (or $W - a = 2B$) and $\psi = b/2$ (or $W - a = B/2$). The plane strain solution is shown by the broken line. It is natural that the three-dimensional solution provides a more accurate limit load for smaller values of $B/(W - a)$ except for small ψ—values when all the solutions predict almost the same result. It is however important to mention that even for

Fig. 6.10 Comparison of the three-dimensional and plane strain solutions

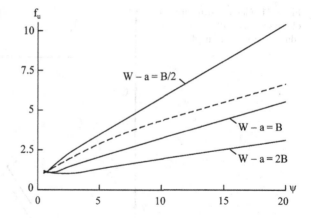

$W - a = B$ the three-dimensional solution gives a much lower value of the upper bound limit load if the value of ψ is large enough.

6.2 Mis-Match Effect

The only material parameter involved in the previous solutions given in this monograph is σ_0. This parameter has no effect of the solutions in dimensionless form (Gibbings 2011). However, in the case of real welded structures there are at least two material parameters that have the dimension of stress, σ_0 and σ_b, where σ_b is the yield stress in tension of base material. Therefore, solutions should depend on the ratio $M = \sigma_b/\sigma_0$ which is usually called the mis-match factor. The definition for the undermatched joints is $M < 1$ and the mis-match factor is much less than 1 in the case of highly undermatched joints. The definition for the overmatched joints is $M > 1$. In the present section, the effect of the mis-match factor on the plastic limit load is demonstrated by means of simple plane strain solutions for both undermatched and overmatched center cracked specimens. The thickness of specimens is $2B$.

6.2.1 Overmatched Center Cracked Specimen

The geometry of the specimen and the orientation of the axes of Cartesian coordinates (x, y) are shown in Fig. 2.1. Upper bound solutions for this specimen have been proposed in Joch et al. (1993) and Alexandrov et al. (1999). Joch et al. (1993) have also given a solution for bending. A solution for scarf-joint specimens has been proposed in Alexandrov and Goldstein (1999).

Fig. 6.11 General structure
of the kinematically
admissible velocity field

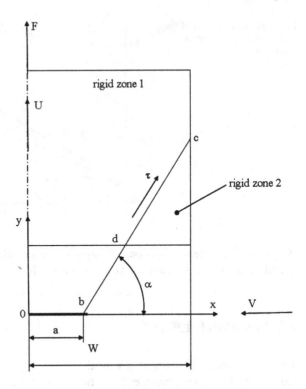

Choose the Cartesian coordinate system such that its axes x and y coincide with
the axes of symmetry of the specimen. Then, it is sufficient to find the solution in
the domain $x \geq 0$ and $y \geq 0$. The natural velocity boundary conditions are

$$u_x = 0 \tag{6.59}$$

at $x = 0$ and

$$u_y = 0 \tag{6.60}$$

at $y = 0$ in the range $a \leq x \leq W$. The general structure of the kinematically
admissible velocity field in one quarter of the specimen is shown in Fig. 6.11.
It consists of two rigid zones separated by the velocity discontinuity line bc.
Its portion bd lies within the weld and portion dc within the base material. Let l_{bc}
be the total length of the velocity discontinuity line and l_{bd} be the length of its
portion within the weld. The length of portion dc is $l_{dc} = l_{bc} - l_{bd}$. It is seen from
Fig. 6.11 that

$$l_{bc} = \frac{W - a}{\cos \alpha}, \quad l_{bd} = \frac{(W - a)h}{\sin \alpha}, \quad l_{dc} = \frac{2(W - a)(\sin \alpha - h \cos \alpha)}{\sin 2\alpha} \tag{6.61}$$

where α is the orientation of the velocity discontinuity line relative to the x-axis and $h = H/(W - a)$. Rigid zone 1 moves along the y-axis with a prescribed velocity U. Rigid zone 2 moves along the x-axis with a velocity V. The magnitude of this velocity should be found from the solution.

The kinematically admissible velocity filed includes no plastic zone of finite size. Therefore, the inequality (1.5) can be used. Taking into account that the specimen is piece-wise homogeneous, this inequality becomes

$$\iint\limits_{S_v} (t_i v_i)dS \le \frac{\sigma_0}{\sqrt{3}} \iint\limits_{S_{dw}} |[u_\tau]|dS + \frac{\sigma_b}{\sqrt{3}} \iint\limits_{S_{db}} |[u_\tau]|dS \qquad (6.62)$$

where S_{dw} is the part of S_d within the weld and S_{db} is the part of S_d within the base material.

The normal velocity must be continuous across the velocity discontinuity line. Therefore, it follows from Fig. 6.11 that

$$V = U \cot \alpha. \qquad (6.63)$$

The velocity components tangent to the velocity discontinuity line are $U \sin \alpha$ and $-V \cos \alpha$ in rigid zones 1 and 2, respectively (the positive direction is indicated by the vector τ in Fig. 6.11). Therefore, the amount of velocity jump across the velocity discontinuity line is $|[u_\tau]| = U \cos \alpha + V \sin \alpha$. Substituting Eq. (6.63) into this equation gives

$$|[u_\tau]| = \frac{U}{\sin \alpha}. \qquad (6.64)$$

Since the amount of velocity jump is constant, the rate of work dissipation at the velocity discontinuity line is simply

$$\sigma_0 \iint\limits_{S_{dw}} |[u_\tau]|dS + \sigma_b \iint\limits_{S_{db}} |[u_\tau]|dS = 2B|[u_\tau]|(\sigma_0 l_{bd} + \sigma_b l_{dc}). \qquad (6.65)$$

The rate at which external forces F do work on one quarter of the specimen is given by Eq. (2.27). Therefore, substituting Eqs. (6.61) and (6.64) into Eq. (6.65) and the result into Eq. (6.62) yield

$$f_u = \frac{F_u}{4BW\sigma_0} = \frac{\lambda(\beta + \tan \alpha)}{\sin^2 \alpha}, \quad \lambda = \frac{(W - a)}{\sqrt{3}MW}, \quad \beta = \frac{H(M - 1)}{W - a}. \qquad (6.66)$$

The magnitude of f_u depends on one parameter, α. Therefore, the right hand side of Eq. (6.66) should be minimized with respect to this parameter to obtain the best upper bound limit load based on the kinematically admissible velocity field chosen. The equation to determine the value of α at which f_u attains a minimum is $df_u/d\alpha = 0$. Then, it follows from Eq. (6.66) that

$$\sin \alpha \cos 2\alpha + 2\beta \cos^3 \alpha = 0. \qquad (6.67)$$

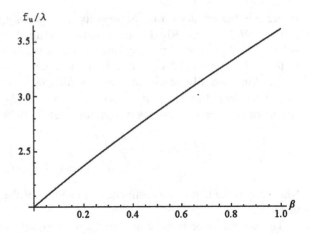

Fig. 6.12 Variation of the dimensionless plastic limit load with β

This equation can be reduced to a cubic equation relative to $\cos \alpha$ (Alexandrov et al. 1999). Therefore, its solution can be found in analytic form, even though the final expression is cumbersome. It is given in Alexandrov et al. (1999). Substituting this value of α into Eq. (6.66) leads to the final expression for the upper bound limit load. An important feature of Eq. (6.67) is that its solution depends on the single parameter β rather than M and geometric parameters separately. In particular, the dimensionless plastic limit load defined by f_u/λ solely depends on β. Its variation with β is illustrated in Fig. 6.12. The dimensionless upper bound limit load found from Eq. (6.66) where α is excluded by means of the solution to Eq. (6.67) will be denoted by $f_u^{(1)}$ for further convenience.

The solution found is not valid for very small and very large cracks. In the case of large cracks, the entire velocity discontinuity line is located within the weld (i.e. this line intersects the line $x = W$ in the range $0 < y \leq H$. In this case $l_{dc} = 0$ and $l_{bd} = (W - a)/\cos \alpha$ (Fig. 6.11). Then, using Eqs. (2.27), (6.64) and (6.65) Eq. (6.62) can be rewritten as

$$f_u = \frac{F_u}{4BW\sigma_0} = \frac{2(1 - a/W)}{\sqrt{3}\sin 2\alpha}. \tag{6.68}$$

It is evident that f_u attains a minimum at $\alpha = \pi/4$. Therefore, the final solution in this case follows from Eq. (6.68) in the form

$$f_u = f_u^{(2)} = \frac{2(1 - a/W)}{\sqrt{3}}. \tag{6.69}$$

In the case of small cracks, no plastic deformation occurs in the weld. The general structure of the kinematically admissible velocity field is shown in Fig. 6.13. It consists of three rigid zones. Rigid zone 1 moves with velocity U in the vertical direction. Rigid zone 3 is motionless. Rigid zone 2 slides along the lower velocity discontinuity line. It is obvious that the limit load in this case coincides with the limit load of the homogeneous specimen with no crack wholly

Fig. 6.13 General structure
of the kinematically
admissible velocity field

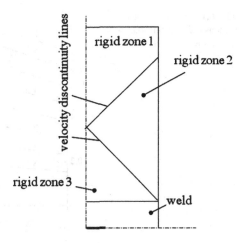

made of base material. Therefore, the solution is (see, for example, Kim and
Schwalbe 2001)

$$f_u = f_u^{(3)} = \frac{2}{\sqrt{3}M}. \tag{6.70}$$

Using Eqs. (6.66), (6.69) and (6.70) the upper bound limit load for a given set of
parameters is determined from

$$f_u = \min\left\{f_u^{(1)}, f_u^{(2)}, f_u^{(3)}\right\}. \tag{6.71}$$

In particular, the value of $f_u = f_u^{(3)}$ should be chosen for the specimen with no
crack independently of other parameters. In general, the value of f_u given by
Eq. (6.71) depends on three parameters, β, a/W and M. Its variation with β at
$M = 1.5$ and several values of a/W is depicted in Fig. 6.14 (curve 1 corresponds
to $a/W = 0.1$, curve 2 to $a/W = 0.2$, curve 3 to $a/W = 0.3$, curve 4 to $a/W =$
0.4, curve 5 to $a/W = 0.5$, curve 6 to $a/W = 0.6$, curve 7 to $a/W = 0.7$, and
curve 8 to $a/W = 0.8$).

6.2.2 Undermatched Center Cracked Specimen

The velocity field shown in Fig. 6.11 is kinematically admissible for undermatched specimens. Therefore, the solution given by Eqs. (6.66) and (6.67) is in
general valid in this case as well. However, specimens with large and small cracks
should be treated differently from the overmatched case. In particular, the solution
(6.70) should be excluded from consideration because the base material is stronger
than the weld. Equation (6.68) is valid when the entire velocity discontinuity line
is within the weld. An unconstrained minimum of the right hand side of the

Fig. 6.14 Variation of the dimensionless limit load with geometric parameters at $M = 1.5$

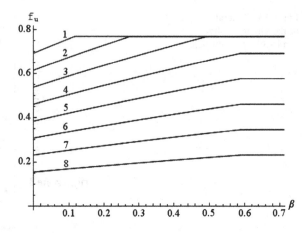

equation is attained at $\alpha = \pi/4$. However, the condition that the entire velocity discontinuity line is within the weld imposes the following restriction on the magnitude of the angle α, $\alpha \leq \alpha_c = \arctan[H/(W - a)]$ (Fig. 6.11). Therefore, Eq. (6.69) should be replaced with

$$f_u = f_u^{(2)} = \begin{cases} \dfrac{2(1 - a/W)}{\sqrt{3}} & \text{for} & \dfrac{H}{W - a} \geq 1 \\[3mm] \dfrac{2(1 - a/W)}{\sqrt{3}\sin 2\alpha_c} & \text{for} & \dfrac{H}{W - a} < 1 \end{cases}. \tag{6.72}$$

In order to clarify why this equation has not been considered for the over-matched case, the function $\Sigma(\alpha, \beta) = (\beta + \tan \alpha)/\sin^2 \alpha$ is depicted in Fig. 6.15 for four values of β. Note that the function $\Sigma(\alpha, \beta)$ is proportional to f_u determined by Eq. (6.66). Curve 1 in Fig. 6.15 corresponds to $\beta = 0.2$, curve 2 to $\beta = -0.1$, and curve 3 to $\beta = -0.4$. The broken curve corresponds to $\beta = \beta_c = -1/(3\sqrt{3})$. It follows from Eq. (6.66) that $\beta > 0$ for $M > 1$ and $\beta < 0$ for $M < 1$. Therefore, curve 1 illustrates the function $\Sigma(\alpha, \beta)$ for the overmatched case and the other curves for the undermatched case. It is seen from Fig. 6.15 that curves 1 and 2 have a local minimum in the range $0 < \alpha < \pi/2$. Curve 3 represents a monotonically increasing function of α. It is possible to show analytically that the broken curve has a point of inflection at $\alpha = \pi/6$. Therefore, Eq. (6.67) has no real root for $\beta < \beta_c$. Moreover, its root at $\beta = \beta_c$ does not correspond to a minimum of f_u. Equation (6.66) should be excluded from consideration for specimens with $\beta \leq \beta_c$. For, f_u decreases as α decreases (Fig. 6.15). However, Eq. (6.66) is not valid for $\alpha < \alpha_c$ since l_{dc} becomes negative and should be replaced with Eq. (6.72). The final expression for the upper bound limit load in the undermatched case is

$$f_u = \left\langle \begin{array}{ll} f_u^{(2)}, & \beta \leq \beta_c \\ \min\{f_u^{(1)}, f_u^{(2)}\}, & \beta > \beta_c \end{array} \right. . \tag{6.73}$$

Fig. 6.15 Illustration of behaviour of the function $\Sigma(\alpha, \beta)$ at several β-values

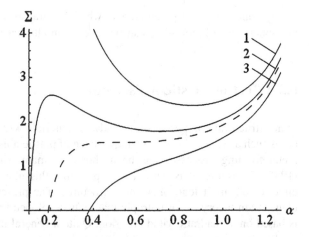

Fig. 6.16 Variation of the dimensionless limit load with geometric parameters at $M = 0.8$

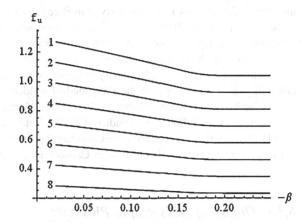

Here $f_u^{(2)}$ is determined by Eq. (6.72) and $f_u^{(1)}$ by Eq. (6.66) assuming that $f_u^{(1)} = f_u$. The variation of f_u determined by Eq. (6.73) with β at $M = 0.8$ and several values of a/W is depicted in Fig. 6.16 (curve 1 corresponds to $a/W = 0.1$, curve 2 to $a/W = 0.2$, curve 3 to $a/W = 0.3$, curve 4 to $a/W = 0.4$, curve 5 to $a/W = 0.5$, curve 6 to $a/W = 0.6$, curve 7 to $a/W = 0.7$, and curve 8 to $a/W = 0.8$).

Note that the solution given by Eq. (6.72) requires that plastic deformation is confined to the weld. This is the definition for highly undermatched specimens. Therefore, the highly undermatched case is obtained when $f_u = f_u^{(2)}$ in Eq. (6.73). The transition between the undermatched and highly undermatched cases is determined by the equation $f_u^{(1)} = f_u^{(2)}$. It is however evident that this transition is conditional, unless the exact value of the limit load solution is available. In fact, choosing another kinematically admissible velocity field in the weld and assuming that the base material is rigid it is possible to find the value of the upper bound limit load different from $f_u^{(2)}$. Comparing this value of the upper bound limit load and $f_u^{(1)}$ another set of parameters at which the transition occurs can be found. In particular, the solution (2.7) can be used as an alternative to the solution (6.72).

6.3 Effect of Plastic Anisotropy

Many structural components are made of materials with strong anisotropic properties such as rolled sheets. The importance of plastic anisotropy on failure in sheet metal forming processes has been shown in many works, for example, Chan (1985). Therefore, it is natural to expect that this material property has a great effect of the limit load of welded structures. The present section deals with the plane strain theory for orthotropic materials developed by Hill (1950). This theory is based on a quadratic yield criterion. Quite a general plane strain yield criterion for homogeneous, incompressible material which complies with the principle of maximum plastic work has been derived in Rice (1973). The solutions given in this section can be generalized to this yield criterion with no difficulty. The specific solutions are given for the center cracked specimen of thickness $2B$ (Fig. 2.1). These solutions have been proposed in Alexandrov and Gracio (2003) and Alexandrov et al. (2007). Other analytic upper bound solutions demonstrating the effect of plastic anisotropy on the limit load of welded structures have been given in Alexandrov and Kontchakova (2004, 2005), Alexandrov and Tzou (2007), Alexandrov et al. (2008), Alexandrov (2010a), and Aleksandrov and Goldstein (2011). A numerical procedure for evaluating the limit load of structures made of anisotropic materials has been used in Capsoni et al. (2001a, b).

6.3.1 Theory of Anisotropic Plasticity

In the case of orthotropic materials the principal axes of anisotropy are orthogonal. The theory used in the present monograph is restricted to a class of anisotropic materials in which the principal axes of anisotropy coincide with the axes of Cartesian coordinates (x, y, z). By assumption, the principal strain rate $\xi_{zz} = 0$. The plane strain orthotropic yield criterion proposed by Hill (1950) is

$$\frac{\left(\sigma_{xx} - \sigma_{yy}\right)^2}{4(1 - c)} + \sigma_{xy}^2 = T^2 \tag{6.74}$$

where σ_{xx}, σ_{yy} and σ_{xy} are the components of the stress tensor in the Cartesian coordinates chosen, T is the shear yield stress in the xy plane, and the parameter c is defined by

$$c = 1 - \frac{1}{4T^2Z^2(KP + PQ + KQ)},$$

$$2K = \frac{1}{Y^2} + \frac{1}{Z^2} - \frac{1}{X^2}, \quad 2P = \frac{1}{Z^2} + \frac{1}{X^2} - \frac{1}{Y^2}, \quad 2Q = \frac{1}{X^2} + \frac{1}{Y^2} - \frac{1}{Z^2}. \tag{6.75}$$

Here X, Y, and Z are the tensile yield stresses in the principal directions of anisotropy, x, y and the thickness direction, respectively. The magnitude of the shear stress referred to the slip lines is given by

$$\sigma_{\alpha\beta} = T\sqrt{1 - c\sin^2 2\phi} \tag{6.76}$$

where ϕ is the anti-clockwise orientation of the slip line relative to the x-axis. In the case of isotropic material $X = Y = Z$. Therefore, it immediately follows from Eq. (6.75) that $c = 0$ and from Eq. (6.76) that $\sigma_{\alpha\beta} = T$. For anisotropic materials, the value of c can vary in the interval $-\infty < c < 1$.

The right hand side of Eq. (1.4) should be rewritten to account for anisotropic properties of material. The first term on the right hand side of Eq. (1.4) is the rate of work dissipation in plastic regions. In order to find this quantity in the case under consideration, it is necessary to supplement the yield criterion (6.74) with its associated flow rule. As a result,

$$\xi_{xx} = \frac{\lambda}{2} \frac{(\sigma_{xx} - \sigma_{yy})}{(1 - c)}, \quad \xi_{yy} = \frac{\lambda}{2} \frac{(\sigma_{yy} - \sigma_{xx})}{(1 - c)}, \quad \xi_{xy} = \lambda\sigma_{xy}, \quad \lambda \geq 0. \tag{6.77}$$

Since $\xi_{zz} = 0$, the equation of incompressibility $\xi_{xx} + \xi_{yy} = 0$ immediately follows from Eq. (6.77). By definition, in the case under consideration the rate of work dissipation in plastic regions is

$$E_V = \iiint\limits_V \left(\sigma_{xx}\xi_{xx} + \sigma_{yy}\xi_{yy} + 2\sigma_{xy}\xi_{xy}\right)dV = \iiint\limits_V \left[(\sigma_{xx} - \sigma_{yy})\xi_{xx} + 2\sigma_{xy}\xi_{xy}\right]dV \tag{6.78}$$

where the equation of incompressibility has been taken into account. Excluding the stress components in Eq. (6.74) using Eq. (6.77) yields

$$\lambda = T^{-1}\sqrt{(1 - c)\xi_{xx}^2 + \xi_{xy}^2}. \tag{6.79}$$

Excluding the stress components in Eq. (6.78) by means of Eq. (6.77) and then excluding λ by means of Eq. (6.79) lead to

$$E_V = 2T \iiint\limits_V \sqrt{(1 - c)\xi_{xx}^2 + \xi_{xy}^2}dV. \tag{6.80}$$

Using Eq. (6.76) the rate of work dissipation at velocity discontinuity surfaces can be written as

$$E_d = T \iint\limits_{S_d} |[u_\tau]| \sqrt{1 - c \sin^2 2\phi} dS.$$ (6.81)

Finally, combining Eqs. (6.80) and (6.81) and replacing the real velocity field with kinematically admissible give

$$\iint\limits_{S_v} (t_i v_i) dS \le 2T \iiint\limits_V \sqrt{(1 - c)\zeta_{xx}^2 + \zeta_{xy}^2} dV + T \iint\limits_{S_d} |[u_\tau]| \sqrt{1 - c \sin^2 2\phi} dS.$$

(6.82)

This equation replaces Eq. (1.4). Analogously, Eq. (1.5) should be replaced with

$$\iint\limits_{S_v} (t_i v_i) dS \le T \iint\limits_{S_d} |[u_\tau]| \sqrt{1 - c \sin^2 2\phi} dS.$$ (6.83)

It is evident that Eqs. (6.82) and (6.83) reduce to Eqs. (1.4) and (1.5), respectively, if $c = 0$ and $T = \sigma_0/\sqrt{3}$.

No material properties other than plastic incompressibility are involved in the definition for kinematically admissible velocity fields. Anisotropic materials considered in the present monograph are supposed to be incompressible. It is therefore obvious that any kinematically admissible velocity field used to get an upper bound solution for a structure made of isotropic materials is also a kinematically admissible velocity field for the same structure made of anisotropic materials. Of course, the velocity boundary conditions must be the same for both structures. Even though no general theory has yet been developed to prove that the asymptotic representation of the equivalent strain rate shown in Eq. (1.9) is valid for anisotropic materials, particular solutions demonstrate that this asymptotic expansion is correct in many cases (Collins and Meguid 1977; Alexandrov 2009, 2010b, 2011).

6.3.2 Overmatched Center Cracked Specimen

It is assumed that the weld is isotropic and the base material obeys the yield criterion (6.74). As before, the tensile yield stress of the weld will be denoted by σ_0. There are two essential parameters which have the dimension of stress, T and σ_0. However, the term "overmatched" used before should be clarified for two reasons. First, the yield criterion of the weld is qualitatively different from that of the base material. The former is the Mises isotropic yield criterion whereas the latter is the quadratic anisotropic yield criterion. Second, the yield stress of the base material depends on the direction of loading. Therefore, in the present section the mis-match factor is defined by $M = \sigma_0/(\sqrt{3}T)$. This definition coincides with the conventional definition for the mis-match factor used in the previous section if both materials are isotropic and obey the same yield criterion. The overmatched specimens are defined by the inequality $M > 1$. Of course, other definitions for the mis-match factor in the case of structures made of a combination of isotropic and anisotropic materials are possible.

The velocity field illustrated in Fig. 6.11 is kinematically admissible for the problem under consideration. Equations (6.61) and (6.64) are valid. Equation (2.27) is valid as well. Since the kinematically admissible velocity field includes no plastic zone of finite size, the upper bound theorem in the form of the inequality (6.83) can be used. It follows from the definition for ϕ and Fig. 6.11 that $\phi = \alpha$. Therefore, the inequality (6.83) can be rewritten as

$$\iint\limits_{S_v} (t_i v_i) dS \le \frac{\sigma_0}{\sqrt{3}} \iint\limits_{S_{dw}} |[u_\tau]| dS + T \iint\limits_{S_{db}} |[u_\tau]| \sqrt{1 - c \sin^2 2\alpha} dS \qquad (6.84)$$

where S_{dw} is the portion of S_d within the weld and S_{db} is the portion of S_d within the base material. Since the amount of velocity jump is constant, the rate of work dissipation at the velocity discontinuity line is determined as

$$\frac{\sigma_0}{\sqrt{3}} \iint\limits_{S_{dw}} |[u_\tau]| dS + T \iint\limits_{S_{db}} |[u_\tau]| \sqrt{1 - c \sin^2 2\alpha} dS$$
$$= \frac{2BTU}{\sin \alpha} \left(l_{bd} M + l_{dc} \sqrt{1 - c \sin^2 2\alpha} \right).$$

Substituting this equation and Eq. (2.27) into Eq. (6.84) and using Eq. (6.61) lead to

$$f_u = \frac{F_u}{4BW\sigma_0} = \frac{(1 - a/W)}{\sqrt{3} M \sin^2 \alpha} \left[hM + (\tan \alpha - h) \sqrt{1 - c \sin^2 2\alpha} \right]. \qquad (6.85)$$

The right hand side of this equation should be minimized with respect to α. It is evident from the structure of Eq. (6.85) that, in contrast to the isotropic case, the value of α at which f_u attains a minimum depends on three dimensionless parameters h, M and c. In general, the solution for f_u depends on the forth parameter, a/W. However, the dimensionless upper bound limit load in the form f_u/λ is independent of a/W (λ has been introduced in Eq. (6.66)). The variation of this load with c at $M = 1.5$ and several h—values is depicted in Fig. 6.17 (curve 1 corresponds to $h = 0.1$, curve 2 to $h = 0.2$ and curve 3 to $h = 0.3$). To further illustrate the effect of plastic anisotropy on the limit load, the variation of f_u/λ with h at several c—values is shown in Fig. 6.18 including the isotropic solution at $c = 0$ (curve 1 corresponds to $c = 0.5$, curve 2 to $c = 0$, curve 3 to $c = -0.5$, curve 4 to $c = -1$, curve 5 to $c = -2$, curve 6 to $c = -3$, and curve 7 to $c = -5$).

Specimens with very small and very large cracks should be treated separately, as in the isotropic case. Details are not given here. However, a necessary condition $l_{dc} \ge 0$ has been verified in course of obtaining the results illustrated in Figs. 6.17 and 6.18.

Fig. 6.17 Variation of the dimensionless upper bound limit load with c at $M = 1.5$

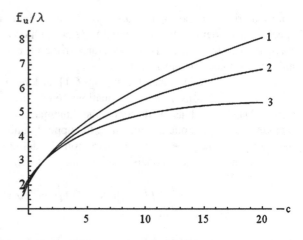

Fig. 6.18 Variation of the dimensionless upper bound limit load with h at $M = 1.5$

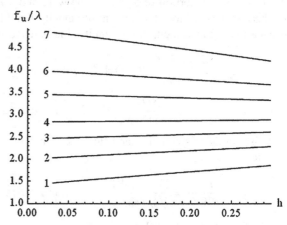

6.3.3 Highly Undermatched Center Cracked Specimen

It is now assumed that the weld is much softer than the base material. The weld material obeys the yield criterion (6.74). The base material is rigid and its velocity is U. The Cartesian coordinate system is chosen as in Sect. 2.1. The velocity boundary conditions are given in Eqs. (2.1) to (2.4). Because of symmetry, it is sufficient to find the solution in the domain $x \geq 0$ and $y \geq 0$.

In order to find an accurate upper bound solution, it is advantageous to adopt the solution for compression of an anisotropic layer between parallel plates proposed in Collins and Meguid (1977). In particular, using the dimensionless quantities introduced in Eq. (2.8) and modifying the velocity field found in this work to be applicable for tension of the layer result in

$$\frac{u_y}{U} = \cos 2\vartheta, \quad \frac{u_x}{U} = -\frac{\varsigma}{h} - 2\sqrt{1 - c}\sin 2\vartheta + u_0 \qquad (6.86)$$

Fig. 6.19 General structure of the kinematically admissible velocity field

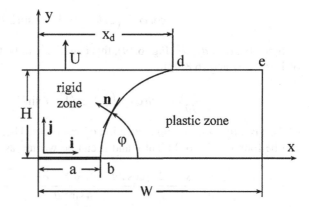

where u_x and u_y are the velocity components in the Cartesian coordinates (x, y) shown in Fig. 6.19, u_0 is an arbitrary constant and

$$\eta = \cos 2\vartheta, \quad d\eta = -2 \sin 2\vartheta d\vartheta. \tag{6.87}$$

It follows from Eqs. (6.86) and (6.87) that the velocity field satisfies the boundary conditions (2.1) and (2.3). However, the boundary condition (2.2) is not satisfied. Therefore, it is necessary to introduce a rigid zone near the axis $x = 0$. The boundary condition (2.4) should be treated in the same manner as for the isotropic case. The rigid zone moves along the y-axis along with the block of rigid base material and the rigid plastic boundary should contain the crack tip. The general structure of the kinematically admissible velocity field chosen is illustrated in Fig. 6.19 (one quarter of the weld is shown). The flow pattern consists of one plastic zone and one rigid zone. There are two velocity discontinuity curves, *bd* and *de*.

Consider the velocity discontinuity curve *bd*. Using Eq. (6.86) the velocity vector in the plastic zone can be represented in the following form

$$\mathbf{u} = -U\left(\frac{\varsigma}{h} + 2\sqrt{1 - c} \sin 2\vartheta - u_0\right)\mathbf{i} + U \cos 2\vartheta \mathbf{j} \tag{6.88}$$

where $\mathbf{i}$ and $\mathbf{j}$ are the base vectors of the Cartesian coordinate system. The velocity vector in the rigid zone is

$$\mathbf{U} = U\mathbf{j}. \tag{6.89}$$

Let φ be the angle between the tangent to the velocity discontinuity curve *bd* and the x-axis, measured anti-clockwise from the axis. Then, the unit normal vector to this curve is given by

$$\mathbf{n} = -\sin \varphi \mathbf{i} + \cos \varphi \mathbf{j}. \tag{6.90}$$

Assume that $\mathbf{U} \equiv \mathbf{u}_1$ and $\mathbf{u} \equiv \mathbf{u}_2$ in Eq. (1.7). Then, it follows from Eqs. (6.88), (6.89) and (6.90) that

$$2 \sin^2 \vartheta \cot \varphi = \left(\varsigma/h + 2\sqrt{1-c} \sin 2\vartheta - u_0 \right). \tag{6.91}$$

Since $\tan \varphi = dy/dx$ (Fig. 6.19), this equation can be transformed, with the use of Eqs. (2.8) and (6.87), to

$$\frac{d\varsigma}{d\vartheta} = -2h \cot \vartheta \left(\frac{\varsigma}{h} + 2\sqrt{1-c} \sin 2\vartheta - u_0 \right). \tag{6.92}$$

This is a linear differential equation of first order. Therefore, its general solution can be found with no difficulty and it can be written as

$$\frac{\varsigma}{h} = \frac{s_1 - 2u_0 \cos 2\vartheta - 4\vartheta\sqrt{1-c} + \sqrt{1-c} \sin 4\vartheta}{4 \sin^2 \vartheta} \tag{6.93}$$

where s_1 is a constant of integration. It follows from Eqs. (2.8) and (6.87) that $\vartheta = \pi/4$ at $\eta = 0$. Then, since the velocity discontinuity curve must contain the crack tip, the boundary condition to Eq. (6.92) is $\varsigma = a/W$ for $\vartheta = \pi/4$. Substituting this condition into the general solution (6.93) determines the constant of integration as

$$s_1 = 2\frac{a}{H} + \pi\sqrt{1-c}. \tag{6.94}$$

Then, the solution (6.93) transforms to

$$\varsigma = \varsigma_{bd}(\vartheta) = h\left[\frac{2(a/H - u_0 \cos 2\vartheta) + \sqrt{1-c}(\pi - 4\vartheta + \sin 4\vartheta)}{4 \sin^2 \vartheta} \right]. \tag{6.95}$$

This equation determines the shape of the velocity discontinuity curve bd. It follows from Eqs. (2.8) and (6.87) that $\vartheta = 0$ at $\eta = 1$ (or $y = H$). Therefore, it is seen from Eq. (6.95) that the velocity discontinuity curve bd has a common point with the line $y = H$ if and only if the numerator of the right hand side of Eq. (6.95) vanishes at $\vartheta = 0$. This is possible if

$$u_0 = \frac{a}{H} + \frac{\pi}{2}\sqrt{1-c}. \tag{6.96}$$

Substituting Eq. (6.96) into Eq. (6.95) leads to

$$\varsigma = \varsigma_{bd}(\vartheta) = \frac{4 \sin^2 \vartheta(a/W) + h\sqrt{1-c}(2\pi \sin^2 \vartheta - 4\vartheta + \sin 4\vartheta)}{4 \sin^2 \vartheta}. \tag{6.97}$$

Applying l'Hospital's rule gives

$$\varsigma_d = \frac{a}{W} + \frac{h\pi\sqrt{1-c}}{2}, \quad x_d = a + \frac{H\pi\sqrt{1-c}}{2} \tag{6.98}$$

where $x_d = \varsigma_b W$ (Fig. 6.19). Using Eqs. (2.8), (6.87) and (6.92) the infinitesimal length element of the velocity discontinuity curve bd can be written in the form

$$dl = \sqrt{dx^2 + dy^2} = H\sqrt{\frac{d\varsigma^2}{h^2} + 4\sin^2 2\vartheta d\vartheta^2} = H\sqrt{\left(\frac{d\varsigma_{bd}}{hd\vartheta}\right)^2 + 4\sin^2 2\vartheta} d\vartheta$$

$$= 2H\cot\vartheta\sqrt{\left(\frac{\varsigma_{bd}(\vartheta)}{h} + 2\sqrt{1-c}\sin 2\vartheta - u_0\right)^2 + 4\sin^4\vartheta} d\vartheta.$$

$$(6.99)$$

Here the function $\varsigma_{bd}(\vartheta)$ should be excluded by means of Eq. (6.97). Substituting Eqs. (6.88) and (6.89) into Eq. (1.8) gives the amount of velocity jump across the velocity discontinuity curve bd in the form

$$\|[u_\tau]\|_{bd} = U\sqrt{\left(\varsigma_{bd}(\vartheta)/h + 2\sqrt{1-c}\sin 2\vartheta - u_0\right)^2 + 4\sin^4\vartheta}. \qquad (6.100)$$

It is seen from Fig. 6.19 and the definition for the angle ϕ given after Eq. (6.76) that $\phi = \varphi$ along the velocity discontinuity curve bd. Therefore, taking into account that $dS = 2Bdl$ and substituting Eqs. (6.99) and (6.100) into Eq. (6.81) yield

$$\frac{E_{bd}}{4BWUT} = hI_1,$$

$$I_1 = \int\limits_0^{\pi/4} \left[\left(\frac{\varsigma_{bd}(\vartheta)}{h} + 2\sqrt{1-c}\sin 2\vartheta - u_0\right)^2 + 4\sin^4\vartheta\right]\cot\vartheta\sqrt{1 - c\sin^2 2\varphi} d\vartheta.$$

$$(6.101)$$

Using trigonometric relations results in

$$\sqrt{1 - c\sin^2 2\varphi} = \sqrt{1 - 4c\tan^2\varphi(1 + \tan^2\varphi)^{-2}}. \qquad (6.102)$$

In this equation, $\tan\varphi$ can be expressed as a function of ϑ by means of Eq. (6.91) in which ς should be replaced with $\varsigma_{bd}(\vartheta)$ according to Eq. (6.97). Then, substituting Eq. (6.102) into Eq. (6.101) and excluding u_0 by means of Eq. (6.96) determine the integrand as a function of ϑ. Integrating should be carried out numerically to evaluate I_1. A small vicinity of $\vartheta = 0$ should be treated separately to facilitate the numerical integration. In particular, the right hand side of Eq. (6.97) reduces to the expression $0/0$ as $\vartheta \to 0$. Expanding it in a series gives

$$\varsigma_{bd}(\vartheta) = a/W + \pi h\sqrt{1-c}/2 - 8h\sqrt{1-c}\vartheta/3 + o(\vartheta), \quad \vartheta \to 0. \qquad (6.103)$$

Also, the integrand in Eq. (6.101) reduces to the expression $0 \cdot \infty$ as $\vartheta \to 0$. Expanding the integrand in a series and integrating lead to

$$\int_0^{\delta} \left[\left(\frac{\varsigma_{bd}(\vartheta)}{h} + 2\sqrt{1-c} \sin 2\vartheta - u_0 \right)^2 + 4\sin^4 \vartheta \right] \cot \vartheta \sqrt{1 - c \sin^2 2\vartheta} \, d\vartheta$$

$$\approx \frac{8(c-1)}{9} \delta^2$$

where $\delta \ll 1$. Using this expression the integral in Eq. (6.101) can be evaluated numerically with no difficulty.

The amount of velocity jump across the velocity discontinuity line $y = H$ in the range $x_d \leq x \leq W$ (Fig. 6.19) is simply equal to $|u_x|$ at $\vartheta = 0$. Then, it follows from Eqs. (6.86), (6.96) and (6.98) that

$$|[u_\tau]|_{de} = U \left[(\varsigma - a/W) h^{-1} - \pi \sqrt{1-c}/2 \right]. \tag{6.104}$$

It has been taken into account here that the right hand side of Eq. (6.104) is a monotonically increasing function of ς and $|[u_\tau]|_{de}$ vanishes at $\varsigma = \varsigma_b$ (or $x = x_b$) as follows from Eq. (6.98). Since $dS = 2Bdx$ and $\phi = 0$ along the line $y = H$, substituting Eq. (6.104) into Eq. (6.81) and using Eqs. (2.8) and (6.98) yield

$$\frac{E_{de}}{4BWUT} = \frac{1}{2} \int_{\varsigma_d}^{1} \left[\left(\varsigma - \frac{a}{W} \right) \frac{1}{h} - \frac{\pi}{2} \sqrt{1-c} \right] d\varsigma = \frac{1}{4h} \left(1 - \frac{a}{W} - \frac{\pi h}{2} \sqrt{1-c} \right)^2. \tag{6.105}$$

Using Eqs. (6.86), (2.8) and (6.87) the strain rate components ζ_{xx} and ζ_{xy} can be found in the form

$$\zeta_{xx} = \frac{\partial u_x}{\partial x} = -\frac{U}{H}, \quad \zeta_{xy} = \frac{1}{2} \left(\frac{\partial u_x}{\partial y} + \frac{\partial u_y}{\partial x} \right) = \frac{U\sqrt{1-c}}{H} \cot 2\vartheta. \tag{6.106}$$

Substituting Eq. (6.106) into Eq. (6.80), taking into account that $dV = 2Bdxdy$ and using Eqs. (2.8) and (6.87) give

$$\frac{E_V}{4BWUT} = 2\sqrt{1-c} \int_0^{\pi/4} \int_{\varsigma_{bd}(\vartheta)}^{1} d\varsigma d\vartheta = 2\sqrt{1-c} I_V, \quad I_V = \frac{\pi}{4} - \int_0^{\pi/4} \varsigma_{bd}(\vartheta) d\vartheta. \tag{6.107}$$

Here the function $\varsigma_{bd}(\vartheta)$ should be excluded by means of Eq. (6.97) and, in the vicinity of $\vartheta = 0$, by means of Eq. (6.103). Then, the integral should be evaluated numerically. Eq. (2.27) is valid. Therefore, the inequality (6.82) can be transformed to $F_u U = 2(E_V + E_{bd} + E_{de})$. Then, using Eqs. (6.107), (6.101) and (6.105) the dimensionless upper bound limit load is represented as

Fig. 6.20 Variation of the dimensionless upper bound limit load with a/W at $h = 0.05$

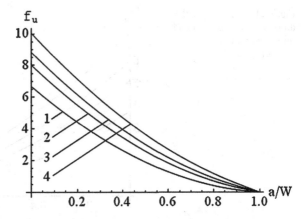

$$f_u = \frac{F_u}{4\sqrt{3}BWT} = \frac{1}{\sqrt{3}}\left[4\sqrt{1-c}I_V + 2hI_1 + \frac{1}{2h}\left(1 - \frac{a}{W} - \frac{\pi h}{2}\sqrt{1-c}\right)^2\right].$$

$$(6.108)$$

The right hand side of this equation contains no free parameters. Therefore, the dimensionless upper bound limit load is determined by numerical integration with no minimization. In the case of isotropic material $\sigma_0 = \sqrt{3}T$. Therefore, the definition for the dimensionless load in Eq. (6.108) coincides with that in Eq. (2.7). It is seen from Fig. 6.19 that this solution is applicable if and only if $x_d \leq W$ ($\varsigma_d \leq 1$). Using Eq. (6.98) this condition can be written in the form

$$\frac{a}{W} + \frac{h\pi\sqrt{1-c}}{2} \leq 1. \qquad (6.109)$$

When the condition (6.109) is not satisfied, the velocity discontinuity curve bd intersects the line $x = W$ (or $\varsigma = 1$) in the range $0 < y < H$ (or $0 < y < H$). Let ϑ_d be the value of ϑ at the point of intersection. Then, it follows from Eq. (6.97) that

$$4\sin^2\vartheta_d\left(1 - \frac{a}{W}\right) = h\sqrt{1-c}(2\pi\sin^2\vartheta_d - 4\vartheta_d + \sin 4\vartheta_d). \qquad (6.110)$$

The integrals involved in Eqs. (6.101) and (6.107) become

$$I_1 = \int_{\vartheta_d}^{\pi/4}\left[\left(\frac{\varsigma_{bd}(\vartheta)}{h} + 2\sqrt{1-c}\sin 2\vartheta - u_0\right)^2 + 4\sin^4\vartheta\right]\cot\vartheta\sqrt{1 - c\sin^2 2\varphi}d\vartheta,$$

$$I_V = \frac{\pi}{4} - \int_{\vartheta_d}^{\pi/4}\varsigma_{bd}(\vartheta)d\vartheta$$

$$(6.111)$$

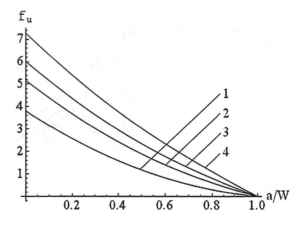

Fig. 6.21 Variation of the dimensionless upper bound limit load with a/W at $h = 0.1$

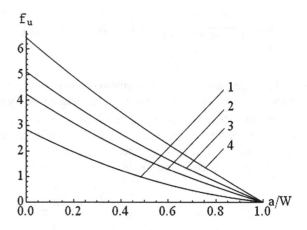

Fig. 6.22 Variation of the dimensionless upper bound limit load with a/W at $h = 0.15$

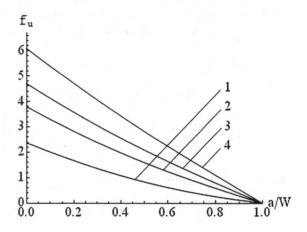

Fig. 6.23 Variation of the dimensionless upper bound limit load with a/W at $h = 0.2$

respectively. There is no velocity discontinuity at $y = H$ in the case under consideration. Therefore, Eq. (6.108) transforms to

$$f_u = \frac{F_u}{4\sqrt{3}BWT} = \frac{\left(4\sqrt{1 - c}I_V + 2hI_1\right)}{\sqrt{3}}. \tag{6.112}$$

Determining ϑ_d from Eq. (6.110), integrating in Eq. (6.111) and substituting the result into Eq. (6.112) give the upper bound limit load when the condition (6.109) is not satisfied.

The general solution depends on three parameters, a/W, h and c. The dependence of the dimensionless upper bound limit load on these parameters found from Eqs. (6.108) and (6.112) is illustrated in Figs. 6.20–6.23 ($h = 0.05$ in Fig. 6.20, $h = 0.1$ in Fig. 6.21, $h = 0.15$ in Fig. 6.22, and $h = 0.2$ in Fig. 6.23). In all these figures, curve 1 corresponds to $c = 0$ (isotropic case), curve 2 to $c = -5$, curve 3 to $c = -10$, and curve 4 to $c = -20$. It is seen from these diagrams that plastic anisotropy has a significant effect of the limit load. Its magnitude is higher for anisotropic material considered in the numerical example. However, it is possible to show the limit load for the isotropic case is higher than that for anisotropic materials if $c > 0$.

References

S.E. Aleksandrov, R.V. Goldstein, Influence of plastic anisotropy on predictions of some engineering approaches in fracture mechanics. Mech. Solids **46**, 856–862 (2011)

S. Alexandrov, A note on the limit load of welded joints of a rectangular cross - section. Fat. Fract. Eng. Mater. Struct. **22**, 449–452 (1999)

S. Alexandrov, Specific features of solving the problem of compression of an orthotropic plastic material between rotating plates. J. Appl. Mech. Technol. Phys. **50**, 886–890 (2009)

S. Alexandrov, Effect of plastic anisotropy on the predictive capacity of flaw assessment procedures. Mater. Sci. Forum **638–642**, 3821–3826 (2010a)

S. Alexandrov An effect of plastic anisotropy on the strain rate intensity factor. in *Proceedings 10th biennial ASME conference on engineering systems design and analysis (ESDA 2010)*, Istanbul (Turkey), 12–14 July 2010, Paper ESDA2010-24021 (2010b)

S. Alexandrov, Behavior of anisotropic plastic solutions in the vicinity of maximum-friction surfaces. J. Appl. Mech. Technol. Phys. **52**, 483–490 (2011)

S. Alexandrov, N. Chikanova, M. Kocak, Analytical yield load solution for overmatched center cracked tension weld specimen. Eng. Fract. Mech. **64**, 383–399 (1999)

S. Alexandrov, K.-H. Chung, K. Chung, Effect of plastic anisotropy of weld on limit load of undermatched middle cracked tension specimens. Fat. Fract. Eng. Mater. Struct. **30**, 333–341 (2007)

S.E. Alexandrov, R.V. Goldstein, An upper bound limit load for overmatched scarf-joint specimens. Fat. Fract. Eng. Mater. Struct. **22**, 975–979 (1999)

S. Alexandrov, Y.-M. Hwang, A limit load solution for highly weld strength undermatched DE(T) specimens. Eng. Fract. Mech. **77**, 2906–2911 (2010)

S. Alexandrov, M. Kocak, Effect of three-dimensional deformation on the limit load of highly weld strength undermatched specimens under tension. Proc. IMechE Part C: J. Mech. Eng. Sci. **222**, 107–115 (2008)

S. Alexandrov, N. Kontchakova, Influence of anisotropy on limit load of weld-strength overmatched cracked plates in pure bending. Mater. Sci. Eng. **387–389A**, 395–398 (2004)

S. Alexandrov, N. Kontchakova, Influence of anisotropy on the limit load of a bi-material welded cracked joints subject to tension. Eng. Fract. Mech. **72**, 151–157 (2005)

S. Alexandrov, G.-Y. Tzou, Influence of plastic anisotropy on limit load of welded joints with cracks. Key Eng. Mater. **345–346**, 425–428 (2007)

S. Alexandrov, G.-Y. Tzou, S.-Y. Hsia, Effect of plastic anisotropy on the limit load of highly undermatched welded specimens in bending. Eng. Fract. Mech. **75**, 3131–3140 (2008)

A. Capsoni, L. Corradi, P. Vena, Limit analysis of orthotropic structures based on Hill's yield condition. Int. J. Solids Struct. **38**, 3945–3963 (2001a)

A. Capsoni, L. Corradi, P. Vena, Limit analysis of anisotropic structures based on the kinematic theorem. Int. J. Plast. **17**, 1531–1549 (2001b)

K.S. Chan, Effects of plastic anisotropy and yield surface shape on sheet metal stretchability. Metall. Trans. **16A**, 629–639 (1985)

J.C. Gibbings, *Dimensional Analysis* (Springer, London, 2011)

I.F. Collins, S.A. Meguid, On the influence of hardening and anisotropy on the plane-strain compression of thin metal strip. Trans. ASME J. Appl. Mech. **44**, 271–278 (1977)

R. Hill, *The Mathematical Theory of Plasticity* (Clarendon Press, Oxford, 1950)

J. Joch, R.A. Ainsworth, T.H. Hyde, Limit load and J-estimates for idealised problems of deeply cracked welded joints in plane-strain bending and tension. Fat. Fract. Eng. Mater. Struct. **16**, 1061–1079 (1993)

Y.-J. Kim, K.-H. Schwalbe, Compendium of yield load solutions for strength mis-matched DE(T), SE(B) and C(T) specimens. Eng. Fract. Mech. **68**, 1137–1151 (2001)

A. Kotousov, M.F.M. Jaffar, Collapse load for a crack in a plate with a mismatched welded joint. Eng. Failure. Anal. **13**, 1065–1075 (2006)

J.R. Rice, Plane strain slip line theory for anisotropic rigid/plastic materials. J. Mech. Phys. Solids **21**, 63–74 (1973)

Index

S. Alexandrov, *Upper Bound Limit Load Solutions for Welded Joints with Cracks*,
SpringerBriefs in Computational Mechanics, DOI: 10.1007/978-3-642-29234-7,
© The Author(s) 2012